天津文化中心 设计卷(下)

Tianjin Cultural Center Design Report（Vol. 2）

本书编委会　编

中国城市出版社
·北京·

图书在版编目 (CIP) 数据

天津文化中心设计卷 . 下 / 《天津文化中心设计卷》编委会编 . — 北京 : 中国城市出版社 , 2012.10
ISBN 978-7-5074-2690-8

Ⅰ . ①天… Ⅱ . ①天… Ⅲ . ①文化中心—建筑设计—设计方案—汇编—天津市 Ⅳ . ① TU242.4

中国版本图书馆 CIP 数据核字 (2012) 第 227076 号

责任编辑	宋 凯（010-63567809 booktaste@163.com）
英文审定	张 垚
英文校对	陈 旭 吴书驰
装帧设计	深圳雅昌设计中心 袁 媛
责任技术编辑	张建军
出版发行	中国城市出版社
地址	北京市西城区广安门南街甲 30 号（邮编：100053）
网址	www.citypress.cn
发行部电话	（010）63454857 63289949
发行部传真	（010）63421417 63400635
总编室电话	（010）68171928
总编室信箱	citypress@sina.com
经销	新华书店
印刷	北京雅昌彩色印刷有限公司
字数	150 千字 印张 31.5
开本	880×1230（毫米） 1/12
版次	2012 年 9 月第 1 版
印次	2012 年 9 月第 1 次印刷
定价	360.00 元

天津文化中心规划设计团队名单（实施方案）

The List of Planning and Design Team for Tianjin Cultural Center（Implementation Scheme）

总体规划设计：
天津市文化中心规划设计组
天津市城市规划设计研究院

大剧院建筑设计（含智能化设计和建筑照明设计）：
冯·格康，玛格及合伙人建筑师事务所(GMP)
现代设计集团华东建筑设计研究院有限公司

博物馆建筑设计（含智能化设计和建筑照明设计）：
华南理工大学建筑设计研究院

美术馆建筑设计（含智能化设计和建筑照明设计）：
KSP尤根·恩格尔建筑师事务所
天津市建筑设计院

图书馆建筑设计（含智能化设计和建筑照明设计）：
株式会社山本理显设计工场
天津市城市规划设计研究院

阳光乐园建筑设计（含智能化设计和建筑照明设计）：
天津华汇工程建筑设计有限公司

市民广场建筑设计（含智能化设计和建筑照明设计）：
美国tvs design设计公司
天津市建筑设计院

管控中心建筑设计：
天津市城市规划设计研究院

General Master Planning Team:
Planning and Design Group of Tianjin Cultural Center
Tianjin Urban Planning & Design Institute

Design Team of Grand Theatre (Including Intelligence and Illumination Design):
von Gerken, Marg and Partners Architects(GMP)
East China Architectural Design & Research Institute Co., Ltd.

Design Team of Museum (Including Intelligence and Illumination Design):
Architectural Design Research Institute of SCUT

Design Team of Art Museum (Including Intelligence and Illumination Design):
KSP Juergen Engel Architeken
Tianjin Architecture Design Institute

Design Team of Library (Including Intelligence and Illumination Design):
Riken Yamamoto & Field Shop
Tianjin Urban Planning & Design Institute

Design Team of Sunshine Paradise (Including Intelligence and Illumination Design):
Huahui Architectural Design & Engineering Co., Ltd.

Design Team of Galaxy mall (Including Intelligence and Illumination Design):
tvs design
Tianjin Architecture Design Institute

Design Team of Managment and Control Center:
Tianjin Urban Planning & Design Institute

环境景观设计:

德国戴水道设计公司

ATLAS（中国）规划 ·设计

天津市城市规划设计研究院

天津市建筑设计院

天津市园林规划设计院

莱茵之华设计集团有限公司（协调配合）

央美光成（北京）建筑设计有限公司（照明设计）

天津市九河市政工程设计咨询有限公司（生态水文施工图设计）

深圳广田装饰集团股份有限公司（标识设计）

地下空间规划与建筑设计:

上海市城市建设设计研究院

天津市城市规划设计研究院

铁道第三勘察设计院集团有限公司

广州地铁设计研究院有限公司

文化中心周边地区城市设计:

SOM

天津市城市规划设计研究院

Design Team of Environmental and Landscape Design:

Atelier Dreiseitl GmbH

ATLAS (CHINA) Planning & Design

Tianjin Urban Planning & Design Institute

Tianjin Architecture Design Institute

Tianjin Institute of Landscape Planning & Design

Rheinschiene GmbH（Coordination and Communication）

Lighting City （Beijing）Architectural Design Co., Ltd（Illumination Design）

Tianjin Jiuhe Municipal Engineering Design Consultation Co., Ltd. (Eco Hydrology Construction Design)

Shenzhen Guangtian Decoration Group Co., Ltd.（Sign Design）

Design Team of Underground Space Planning and Design:

Shanghai Urban Constuction Design Research Institute

Tianjin Urban Planning & Design Institute

The Third Railway Survey and Design Institute Group Corporation

Guangzhou Metro Design & Research Institute Co., Ltd.

Planning Team of Cultural Center Surrounding Area Urban Design:

Skidmore, Owings & Merrill（SOM）

Tianjin Urban Planning & Design Institute

目录 Contents

Contents

1

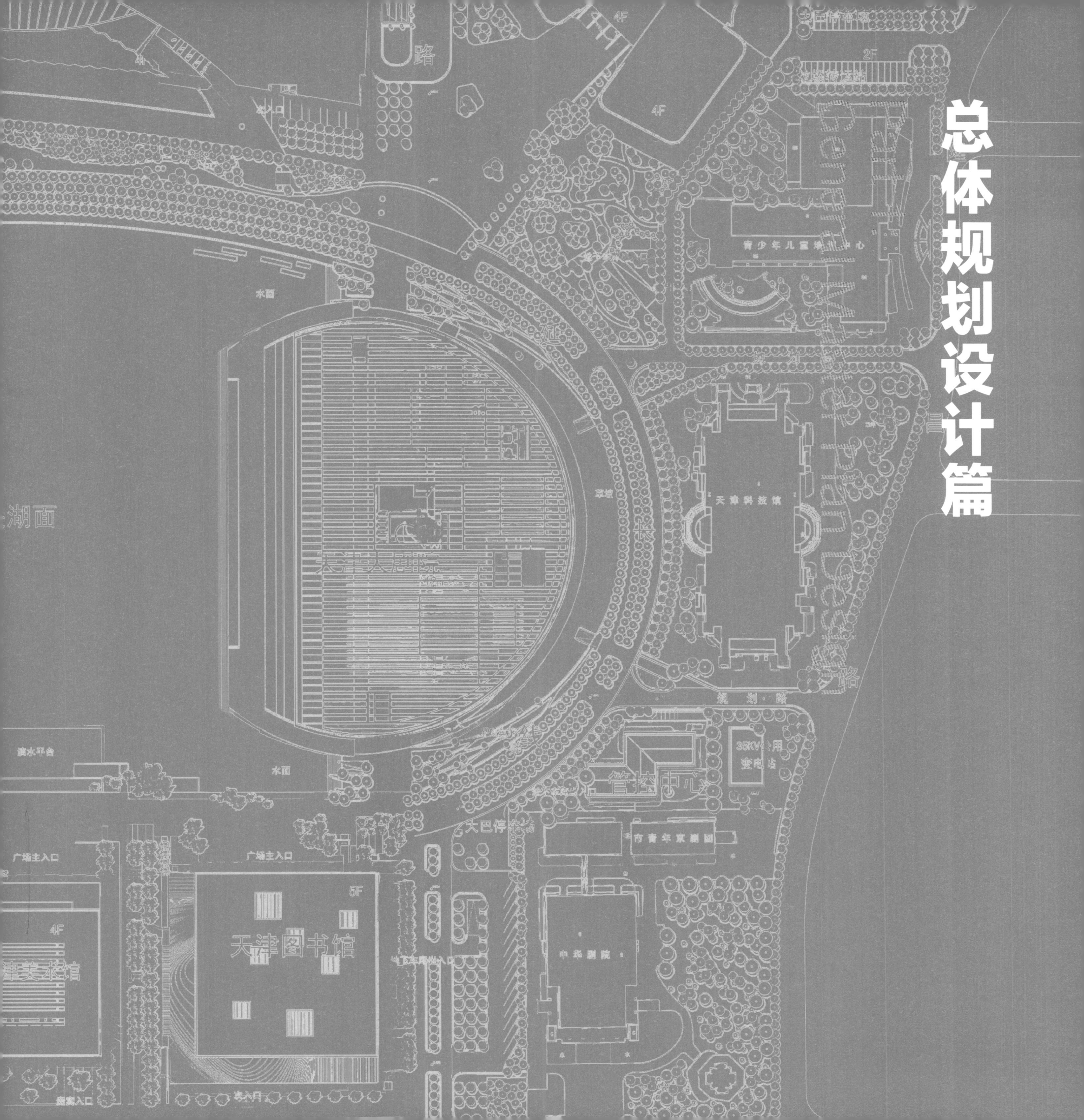

总体规划设计篇

Part I General Master Plan Design

总体规划设计
General Master Plan Design

天津市文化中心规划设计组
天津市城市规划设计研究院
Planning and Design Group of Tianjin Cultural Center
Tianjin Urban Planning & Design Institute

2012年5月，天津文化中心开始向公众开放。这座天津有史以来最大的公共文化设施建设工程，人民的“城市客厅”，终于揭开面纱，张开怀抱，向期待已久的广大市民呈现丰富典雅的文化盛宴。历时四年的规划建设，倾注了各方的心血与汗水、智慧与激情，承载全市人民厚望的文化中心，将成为完善城市功能、弘扬城市文化、提升城市面貌、促使城市发展的强有力城市“心脏”。

In May 2012, Tianjin Cultural Center opened to the public. As the biggest construction project of public cultural facilities in the history of Tianjin and the people's "city living room", Tianjin Cultural Center finally unveiled its mask, opened its arms, and showed a rich and elegant cultural feast to the long-awaited citizens. Taking four years to plan and construct, Tianjin Cultural Center converges the painstaking efforts and endeavors, wisdom and passion from all walks of life. Carrying the earnest expectation of all the citizens, Tianjin Cultural Center will become a strong heart of the city to improve its functions, to promote its culture, to enhance its image and to stimulate its development.

摄影：刘东
Photographer：Liu Dong
英文翻译：窦强
English Interpret：Dou Qiang

“山、水、塔”的中国园林布局
Chinese Garden Layout "Hill, Water and Tower"

城市功能的定位

在中国由南而北，从珠三角、长三角再到京津冀逐级发展的战略中，以及京津同城错位发展的背景下，天津把握前所未有的发展机遇，正在成为带动中国经济发展的第三级和北方的经济中心。2006～2011年，天津累计实施总投资2.27万亿元、共1440项的高水平大项目好项目，进一步优化升级经济结构，形成了航空航天、石油化工、电子信息、生物医药等八大优势支柱产业。五年来，全市生产总值年均增长16.5%，至11年底已达1.12万亿元。经济高速发展的同时，天津更加需要加强文化建设，更加需要凝聚力量，为城市的发展挖掘潜力、拓展空间、增加后劲。纵观天津六百多年沧桑与辉煌的历史，其城市空间形态从设卫开始的老城厢，到逐水而兴的海河沿岸租界地，再到集中延展扩张的中心城区，城市一直围绕着功能性需求的商业商贸中心在发展演变，但中心城区的城市空间一直缺乏具有标志性和凝聚力的城市中心。随着天津经济社会发展取得显著成效，为体现城市发展定位，完善城市文化服务功能，提升城市形象与活力，2008年天津市委、市政府决定规划建设天津文化中心。

Orientation of Urban Function

In China from south to north, from the step by step development strategy of the Pearl River Delta, Yangtze River Delta to Beijing-Tianjin-Hebei region, and under the background of the dislocation development of Beijing and Tianjin city, Tianjin grasps the unprecedented development opportunity and becomes the third level driving Chinese economic development and the northern economic center. From year 2006 to 2011, Tianjin cumulatively implemented a total investment of 2.27 trillion Yuan and a total of 1,440 high-level large and good projects, further optimized and upgraded its economic structure, forming eight advantage pillar industry, such as aerospace, petrochemical industry, electronic information, biological medicine and so on. In five years, the city's annual GDP growth rate is 16.5%, and the city's GDP reaches 1.12 trillion Yuan by the end of 2011. During the rapid economic development, Tianjin is more necessary to strengthen its cultural construction, and more need to consolidate forces to explore the potential of the city's development, expand space and increase stamina. Taking an overall view of the city of Tianjin which has experienced its vicissitudes and glory for six hundred years, its urban spatial form has transformed from the old city area founded on fortification to the concession territory prospered along the Haihe River, then to the intensively extending and expanding central city. The evolution process of the city has been continuously around the functional need of a business and trade center, but the urban space of the central city has always been lack of an iconic and cohesive centrality. As Tianjin's economic and social development has achieved remarkable effects, in order to reflect the development orientation of the city, to improve its cultural service function, and to promote its city image, in 2008, Tianjin Municipal Party Committee and Tianjin Municipal Government decided to carry out the planning and construction of Tianjin Cultural Center.

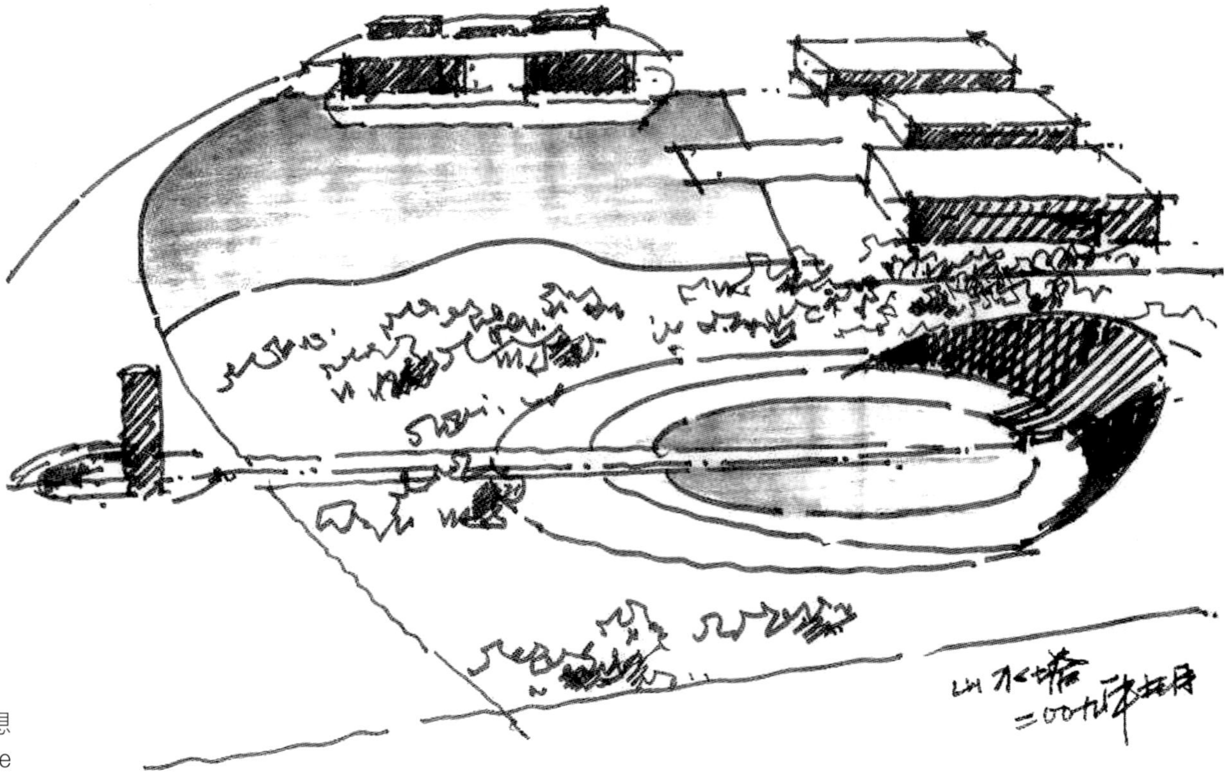

文化中心设计构想
Design Concept for Culture Centre

总平面　Master Plan

基地历史变迁（1998－2012年）
Historcal Site Transition 1998 - 2012

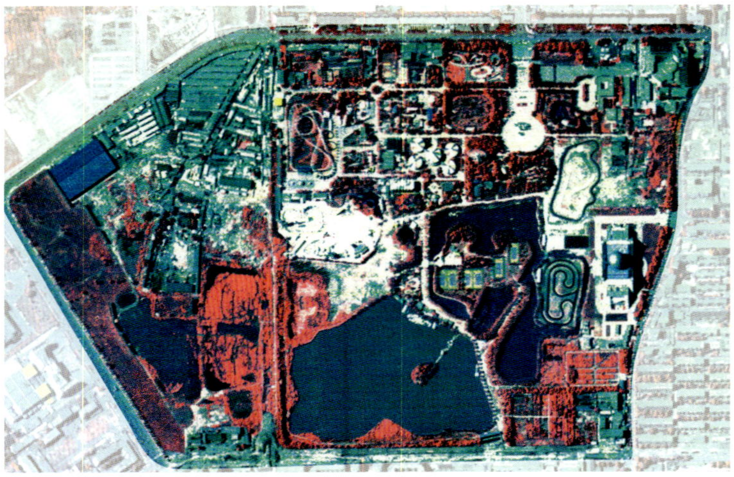

1998

2002

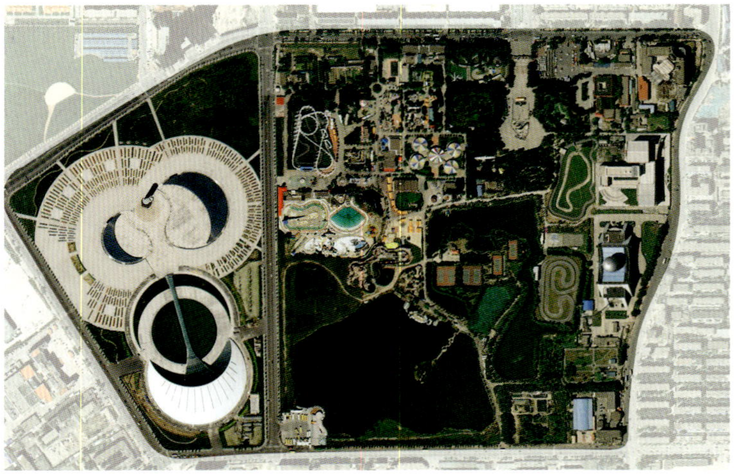

2004

2006

2008

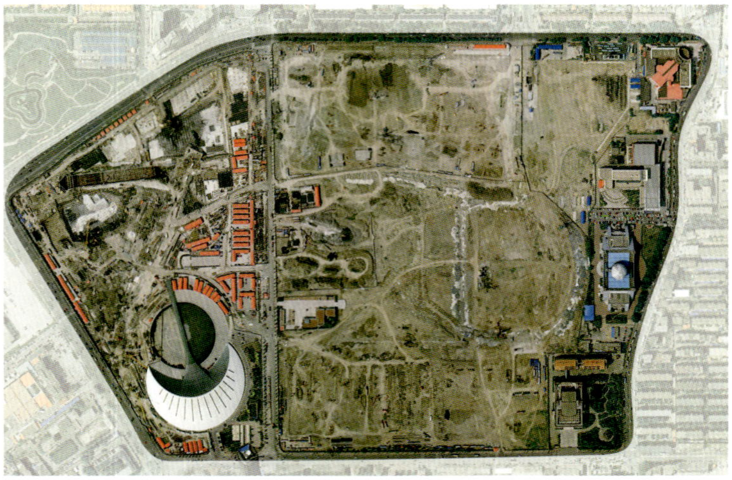

2010

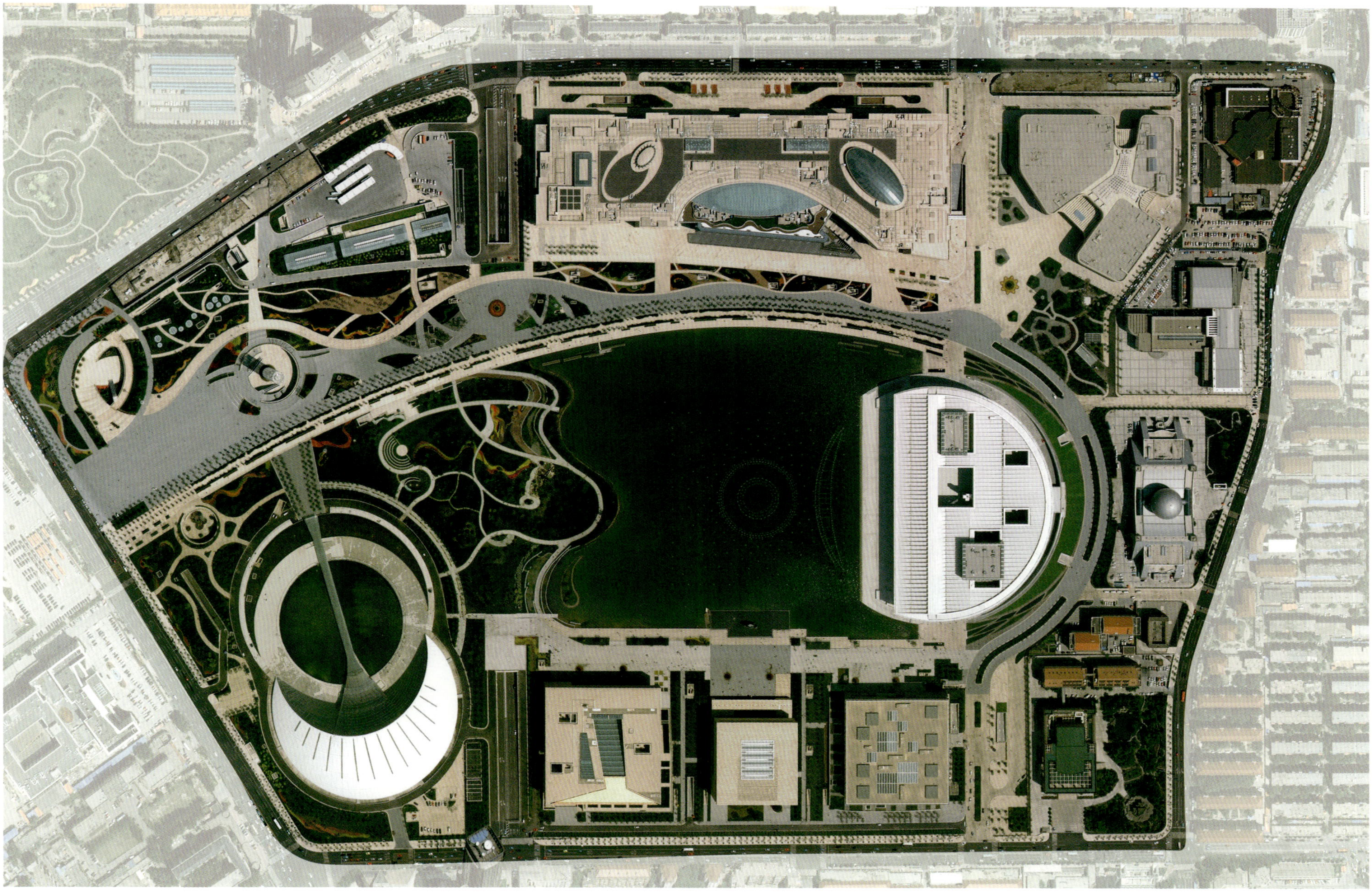

2012

天津文化中心选址于城市中心南部，位于原有城市主中心——小白楼商业商务区南侧约2.4公里处。由于开发趋于饱和，以及临近历史街区、功能组成相对单一的原因，小白楼地区缺乏足够的发展空间与凝聚力。文化设施集聚、功能复合、空间舒展的文化中心已经成为新的城市“心脏”。这是一个高雅而又亲民的城市中心。在这里，“城市客厅”的公共性被突显。由市委、市政府建筑群和大礼堂组成的行政中心居于一侧，把最重要的位置让出来设置文化中心。其中，大剧院作为主导建筑居于中央，与南侧的文化建筑、北侧的商业建筑形成整体和谐的高品质建筑群体，结合绿色、活力的开放空间，为市民提供尽可能多的、典雅亲和的公共活动场所。4条轨道线与公交首末站、大巴停车场以及公共停车场，形成了综合交通枢纽，让广大市民能够便捷地到达文化中心，使其成为真正为市民服务的“市民客厅”。这是一个既能传递现代精神又能传承传统文化的城市中心。规划糅合“山、水、塔”的中国园林布局，以中心湖为核心，以生态岛为山，结合迎宾塔和周边文化建筑，形成了中国山水意境的园林布局。中央湖面结合大型艺术喷泉，形成了优雅而生动的水景主题，隐喻天津特色的“水文化”。同时借鉴西方园林的“大轴线、林荫道”手法，以海棠步道连接大剧院和大礼堂。在海棠花盛开的时节，文化中心将成为天津的“海棠心”。这是一个既能代表天津形象又能凝聚天津力量的城市中心。它以强大的凝聚力，和行政中心、接待中心共同带动文化中心周边地区，形成以文化为主导的，集商务、金融等功能于一体的，引领中心城区发展的文化商务核心区。

地铁规划平面图
Plan of Subway

Tianjin Cultural Center is located in the southern part of the city center, and it is about 2.4 kilometers away from the original main city center – Xiao Bai Lou commercial business district – on the south side. Due to the saturation of development and the relatively singular function organization of nearby historic district, Xiao Bai Lou area lacks enough development space and cohesion. The cultural center, which has integrated cultural facilities, compound functions and extended space, has become a new heart of the city. This is an elegant city center which is close to the people. The public nature of a "city living room" is highlighted here. The administrative center, which is composed of the Municipal Party Committee and Municipal Government building group and the Grand Auditorium, locates on one side, giving the most important position to set the cultural center. Wherein, the Grand Theatre as the dominant building sits in the central area, forming an integrated and harmonious high-quality building group with the cultural buildings on the south side and the commercial buildings on the north side. Combined with green and vibrant open space, elegant and amiable public spaces are provided for the citizens as much as possible. Four subway lines, bus terminals, bus parking lot and public parking lot together form an integrated transport hub, so that the general public can easily reach the cultural center, making it a truly "public living room" serving the public. This is a city center which can not only pass the modern spirit but also inherit the traditional culture. The general plan blends the Chinese garden layout of "hill, water, tower", setting the central lake as the core and the ecological island as the hill, in combination with the Welcome Tower and surrounding cultural buildings forming a garden layout with the spirit of Chinese landscape . The central lake combined with a large-scale artistic fountain forms an elegant and vivid waterscape theme, providing a metaphor of Tianjin's "water culture". At the same time, by drawing lessons from the "large axis, boulevards" of western garden, the Grand Theatre and Grand Auditorium are connected through a begonia promenade. During the blooming season of begonia flowers, the cultural center will become a "begonia heart" of Tianjin. This is a city center which can not only represent the image of Tianjin but also gather the strength of Tianjin. With its strong power of cohesion and together with the administrative center of the reception center, the surrounding area of the cultural center will boost, forming a cultural business core area which is led by culture and integrated with business, financial, and other functions, and which can lead the development of the central city area.

文化中心周边地区
Surrounding Area of Tianjin Cultural Center

景观湖实景
Artificial Lake View

城市建筑的追求

（1）立足文化的城市建筑特色

当下的中国，如何克服“千城一面”，形成地域性城市建筑特色，是各个城市需要直面的课题。建筑是城市文化的表征，建筑特色孕育着城市文化的记忆。建筑只有扎根于地方文化，才能够形成具有内涵的地方特色。

天津是国家历史文化名城。六百多年的发展，形成了深厚的建筑文化积淀。不同时期形成的一些重要的标志性区域，如老城厢、五大道、劝业场、和平路、小白楼、解放北路等等，显露出不同特色。它们各具韵味又统一和谐，形成了多元共存、兼收并蓄的城市风貌。城市的建筑特色既能通过留存下来的建筑遗产表现出来，也能通过反映时代感的新建筑所传承的文化基因展现出来。当代天津致力于传承传统建筑文化，吸收世界文化的优秀元素，以古典建筑为精华、以现代建筑为主导的，彰显整体多元融合、个体庄重大气、中西合璧、古今交融的城市建筑特色。具体到单体建筑特色，在规划管理中采取了分区域、多元化的建筑特色引导。既有历史风貌区的中、西式古典特色建筑，如津湾广场、意式风情区，又有其他区域现代简洁的建筑特色，如津塔、梅江会展中心。天津文化中心尝试在现代建筑共性的基础上，探索既体现时代感又强调地域性的建筑特色。大剧院以现代感的建筑细部传达了中式飞椽重檐的意向，并借鉴中国古建三段式的经典手法，形成大气典雅的殿堂空间，统领整个文化中心。博物馆以“世纪之窗”的概念为设计原点，以“六层重门”、“时光隧道”，回顾天津建城600年的文明，再现近代百年看中华的历史，形成了极具天津特色的展览场所。市民广场以经典的天津“水文化”为灵魂，形成内外一体、动静相宜、平衡和谐的商业空间。简洁洗练、沉稳庄重、新而不怪的建筑形式，完整、明晰的空间构成，适宜的结构体系，精心雕琢的细部，以及整体与个性的平衡，以及地域特色的传承，共同构成了天津文化中心现代经典而又富有文化内涵的建筑特色。

Pursuit of Urban Architecture

(1) The Characteristics of Urban Architecture Based on Culture

Nowadays in China, how to overcome making cities with similar appearance and to form regional characteristics of urban architecture is the task facing each city. Architecture is the representation of urban culture, while architectural characteristics contain the memory of urban culture. Local characteristics with their cultural connotation can only be formed when architecture is rooted in local culture.

Tianjin is a national historical and cultural city. More than 600 years of development, the city has formed a strong architectural culture heritage. Some important and iconic areas formed in different periods, such as the old city area, five avenues, the Quanye bazaar, Heping Road, Xiao Bai Lou, North Jiefang Road, and so on, manifest different characteristics. Each of these areas has its own charm while they are a harmonious unity, forming an eclectic city style of multicultural coexistence. Architectural characteristics of a city can be manifested both through the surviving architectural heritage and through inherited cultural genes of new architecture reflecting the era. Contemporary Tianjin commits to inherit traditional architectural culture, while absorb the outstanding elements of world culture. The characteristics of its urban architecture will take classical architecture as the essence and modern architecture as the dominance, and will highlight the overall multiple fusion, individual gravity and generosity, the combination of the east and the west, the fusion of the past and present. Specific to individual building characteristics, diversified guidance of architectural characteristics subject to specific region was adopted in planning management. There are not only Chinese and Western classical architectural characteristics of historical areas, such as Jin Wan Plaza, Italian Landscape Zone, but also modern concise architectural characteristics of other areas, such as Tianjin Tower, Meijiang Convention and Exhibition Center. On the common basis of modern architecture, Tianjin Cultural Center tries to explore architectural characteristics both reflecting the sense of the times and emphasizing regional characteristics. Via modern architectural details, the Grand Theater conveys the intention of Chinese fly rafters and multiple eaves, and by drawing lessons from the classical principle of tripartite composition in Chinese ancient buildings, the Grand Theater creates grand and elegant hall spaces commanding the whole cultural center. The Museum takes the concept "century window" as its design origin. Through "six layers of heavy doors" and "time tunnel", 600 years of civilization of Tianjin city is given a review and a century of modern Chinese history is represented, forming an exhibition place of strong Tianjin characteristics. The Galary Mall, which takes the classic "water culture" of Tianjin as its soul, creates a balanced and harmonious commercial space integrating inside and outside and keeping balance between action and silence. Concise and sedate new form, complete and clear space composition, suitable structure system, elaborate details and the balance between the whole and individual, as well as the inheritance of regional characteristics, all together constitute the architectural characteristics of Tianjin Cultural Center, which are both modern and classic and which have rich cultural connotations.

文化中心大剧院
Grand Theatre

（2）突出整体的城市设计把握

近年来，天津运用城市设计手段，在规划管理、建设项目管理、历史名城保护、城市环境提升等工作中，强调从整体性上对城市形象和空间形态进行控制和引导，取得了一定的效果。在文化中心，规划设计组注重总体把控，积极发挥统筹协调作用，将整体的城市设计理念贯彻到各阶段规划设计工作中。通过多种形式，建立开放的工作机制与交流平台，实现规划师、建筑师、景观师、工程师、艺术家之间横向协同工作，实现整体最优的目标。同时，加强纵向的整体把控，建立中选方案设计单位负责制，将工作责任范围从方案延续施工图、室内设计直至家具设计并贯穿建造全过程。以南侧文化建筑为例，为了加强南侧文化公园的空间完整性，形成博物馆、美术馆、图书馆三个建筑之间一体化的入口前空间，对美术馆征集方案进行了调整，将向北延伸到湖面的雕塑庭院适当后退，把让出来的空间开放为雕塑广场。在注重整体空间的组织的同时，也加强了对建筑形式的统筹，对建筑体量、材质、色彩、退线、贴线等提出控制与引导。在方案征集阶段，图书馆外墙主要采用玻璃幕墙，美术馆为石材，博物馆则为深色金属。为强化建筑的稳重感与文化品质，利于维护，决定三座建筑统一采用石材外墙，通过不同肌理、材质搭配突出各自个性：博物馆采用石材与铜板搭配，突出厚重感、历史感；美术馆则通过精雕细刻的洞石表现艺术的精致；图书馆结合功能特点采用轻盈透光的石材百叶。

在整体城市设计理念的指导下，以“山、水、塔”布局和“大轴线、林荫道”手法为统领，构建了整体协调的空间格局、和而不同的建筑形态，形成了建筑与空间相互交融的高品质环境，在城市的核心区域营造了和谐共鸣、精致典雅的标志性公共空间。

（2）Urban Design Control Highlighting the Whole

In recent years, by using urban design approach, the control and guidance of Tianjin's city image and spatial morphology in terms of integrity were emphasized in the works of planning management, construction project management, the protection of historic city, the improvement of city environment and so on. For Tianjin Cultural Center, planning design team pays attention to the overall control, plays an active co-ordination role, and implements overall urban design ideas in the various stages of planning and design work. Through various forms, it is tried to establish an open working mechanism and exchange platform, and to realize the horizontal cooperative work among planners, architects, landscape architects, engineers and artists, in order to achieve the overall optimal targets. At the same time, it is tried to strengthen the longitudinal overall control by establishing a responsibility system of selected design firms, which will extend the range of their work responsibilities from design scheme to construction drawing, from interior design to furniture design, and will run through the whole process of the construction. Taking the cultural buildings on the south side of Tianjin Cultural Center as an example, the selected scheme of the art museum was adjusted in order to strengthen the spatial integrity of the cultural park on the south side and to form an integrated space in front of the entrances of the museum, the art museum and the library. The sculpture yard extending to the lake was set backwards properly, and the space let out was open as a sculpture square. With the consideration on the organization of whole space, the co-ordination of architectural form was also strengthened. Control and guidance were given in terms of building volume, material, color, building setback and building near-line. In the scheme soliciting stage, the exterior wall of the library mainly adopted glass curtain wall, the art museum adopted stone, and the museum adopted dark metal. To enhance the sense of stable and cultural character and to be conducive to the maintenance, it was decided that these three buildings use the same stone exterior wall, while the individual character of each building was highlighted via the combination of different textures and materials: the museum using the combination of stone and copper to highlight the sense of heavy and history; the art museum using the exquisitely carved travertine to represent the delicacy of art; combined with functional characteristics, the library using light and opaque stone blind.

Under the guidance of the overall urban design ideas and by using "hill, water, tower" and "large axes and boulevard" as a guide, spatial layout with an overall coordination and building form of harmony but not sameness were produced; a high quality environment which blends architecture and space was formed; and a harmonious, resonant, refined and elegant iconic public space was created in the core area of the city.

湖南岸：文化展示区
South of the Lake: Cultural Exhibition District

湖北岸：商业娱乐区
North of the Lake: Commercial and Recreational District

（3）强调理性的建筑设计理念

文化中心建筑设计坚持理性思考与科学判断，始终关注功能、空间、技术等本质问题，在合理的前提下追求创新，追求功能、空间与形式的均衡，努力营造经济实用、舒适愉悦的整体环境。

通过总结和比对国内外20余个著名城市中心与30余个世界一流文化场馆，在功能定位、环境整体品质、建筑品质、展示收藏、视听效果、照明效果、数字化管理以及公共服务功能等方面建立了108个指标为基础的指标比对体系。以此为参照，完善设计方案，全面提升了设计的成熟度。如，在大剧院设置了多功能厅以及教室，专门作为向市民进行艺术普及教育的公益性课堂，在博物馆设置了母婴室、观众参与室、虚拟展示厅等功能等等，提升了建筑的公益性、专业性，使市民能够更加充分地享受历史、文化与艺术的熏陶。在大剧院中，对视线、声学指标等等进行了研究。其中，歌剧厅内视距小于33m的坐席超过了93%，最远点视距小于34.6m。图书馆在完善传统借阅功能的基础上，升级了古籍修复与数字资源加工功能，设置了专门的音乐图书馆，增加了24小时借阅功能和RFID自主借还书系统，真正成为市民享受知识的“智慧乐园”。

各建筑设计在保证功能合理的前提下，依据各自使用特点形成了内涵丰富、个性鲜明的空间。居于中央的大剧院在亲水台阶上以三个独立的体量分别布置三个艺术场馆，在半圆形屋盖的统一下形成延续到湖面的城市灰空间，既提升了大剧院的城市功能，突出了建筑在文化中心的主体地位，又降低了运营成本。南侧的博物馆、美术馆、图书馆均有源自功能、契合环境的主题空间与建筑形式。北侧的市民广场将“水面”成功引入到内部，以“日、月、星、辰”形成与湖面互动的高品质商业空间。偏居一隅的阳光乐园以“三叶草”作为功能与空间的结合点，赋予建筑以活泼、灵性。

（3）Architectural Design Ideas Emphasizing Rationality

The architectural design of Tianjin Cultural Center adheres to rational thinking and scientific judgment. It has always concerned about fundamental questions such as function, space, technology and so on. In the premise of rationality, it pursuits innovation and the balance between function, space and form, and it endeavors to create an economical, practical, and comfortable environment.

Through summing up and comparing more than 20 domestic and foreign famous city centers and more than 30 world class cultural venues, an index matching system was established based on 108 indicators in terms of function orientation, overall environment quality, architectural quality, display collections, audio-visual effects, lighting effects, digital management and public service functions and so on. By using this index system, design schemes were improved and design maturity was raised overall. For example, the Grand Theatre provides multi-purpose halls and classrooms, becoming a public welfare classroom for carrying out universal art education to the public. The museum provides infant rooms, rooms for audience participation, a virtual exhibition hall and many other functions, which can enhance public welfare and the professional nature of the buildings, so that the audience can more fully enjoy the nurture of history, culture and art. The sight and acoustic indicators of the Grand Theatre has been studied carefully. In the opera chamber, the seats which have a viewing distance less than 33 meters are more than 93%, the viewing distance of the farthest point is less than 34.6 meters. With the improvement of traditional library functions, the library upgraded the functions of ancient books repair and digital resources processing, set up a special music library, and added 24 hours lending function and independent RFID（Radio Frequency Identification）book circulation system, truly becoming a "wisdom paradise" for the public to enjoy knowledge.

The architectural design of each building, in the premise of assuring reasonable functions, forms space of rich connotation and distinct characteristics. The central grand theatre has three separate volumes on the steps near water which are respectively arranged as three art venues. Under the cover of a semi-circular roof, a transitional urban space is created and extended to the lake, which can not only reduce the operation cost, but also enhance the urban function of the theater and highlight the major role of the building with respect to the culture center. Located on the south side of the lake, the museum, art museum and library all have their own themes of space and architectural form derived from function and corresponding to the context. Located on the north side of the lake, Civic Square successfully gets the water of the lake into its interior, forming a high quality commercial space of interaction with the lake by using the theme of "sun, moon, star, earth". At the corner of the site, Sunny Paradise uses "clover" as the point of the integration of function and form, making the building lively and spiritual.

南岸
South View

城市发展的永续

2006年7月，在国务院批复的《天津市城市总体规划》中指出，要努力把天津建设成为生态城市。围绕城市定位，天津正在加快生态城市建设步伐，努力在资源节约、环境保护上走在全国前列。中新生态城以及子牙循环经济产业区等项目的建设，为城市可持续发展提供了宝贵的实践经验和学习样板。

天津文化中心积极采取可持续策略，因地制宜采用生态技术以及各种成熟的综合技术，为项目品质提供充足的技术保障。一、能源站建设：依据不同业态特征、管理权属，文化中心区域集中设置了三处能源站。通过集中建设、集中管理，降低了能源系统的初期投入与运行成本，并采用可再生能源技术（带调峰复合式三工况埋管地源热泵系统）提供冷热源，实现了节能降耗的目标。同时通过减少非空气途径排热，最大限度减少冷却塔或风冷室外机的数量，降低了对环境的压力，美化了地面景观。据测算，预计实际节能效果可达到：每年节约标准煤8873吨，减少CO2排放23249吨、减少SO2排放213吨、减少NOX排放81吨，节水、节能率为36.02%。二、雨水收集处理，湖水循环净化处理：合理组织雨水收集系统、调蓄系统，利用蓄水技术削减地表径流峰值，通过过滤净化流程将雨水汇入中心湖，全年可利用雨水9万立方米。在生态岛南北两侧设有约2600平方米的生态净化群落,通过物理过滤、滤料基质吸附、水生植物根区分解吸收等手段净化湖水。通过整个系统的建设，节约了市政雨水管网升级改造费用约2亿元。每年可节水130万元，节约径流污染整治投资费19万元，节约防洪费用1000万元。三、光伏发电系统：结合市民广场建筑屋面设置，采用专电专用的即发即用控制技术，提高利用效率，节约投资及后期维护成本。文化中心的建设表达了对生态文明的向往，为天津生态城市建设发挥了示范作用

愿天津文化中心成为一个高雅艺术的展示中心、文化艺术的普及中心，一座繁荣恢弘的“人民殿堂”，更是提升城市载体功能、带动城市发展、造福子孙后代的市民乐园。

Continuity of Urban Development

In July 2006, "The General Planning of Tianjin City" which was approved by the State Council, pointed out that we should make great efforts to build Tianjin as an eco-city. According to the city development orientation, Tianjin is accelerating the pace of building an eco-city, and keeping ahead in the country for resource conservation and environment protection. The construction of Zhongxin eco-city, Ziya circular economy industrial area and other projects provided precious practical experiences and learning models for the sustainable development of the city. Sustainable strategies were actively adopted in Tianjin Cultural Center, and ecological technologies with other mature integrated technologies were applied according to local conditions, which provide an adequate technical support for project quality. First, the construction of energy stations: three energy stations were built within the site of the culture centre according to the features of different operational types and management authority. Through concentrated construction and management, the initial investment and operation cost of the energy system was reduced. And by using renewable energy technologies (combined three-condition underground heat pump system with peak adjustment and buried tubes) to provide heat and cold sources, the goal of energy saving has been achieved. At the same time, by reducing heat emission via non-air way, the numbers of cooling tower or outdoor air cooling machine were reduced to to the maximum, so the pressure on environment was lowered and the ground landscape was beautified. According to calculation, it is expected that the actual effect of energy saving can be: the annual savings of 8873 tons of standard coal, the reduction of 23249 tons CO2 emission, 213 tons SO2 emission and 81 tons NOX emission, and a water and energy saving rate of 36.02%. Second, rainwater collection and treatment, lake water circulation and purification treatment: rational organization of rainwater collection system and storage system; the use of water storage technology to reduce surface runoff peak; through filtering and purifying process, rainwater flows into the central lake, which can use 90,000 cubic meters of rainwater all the year around. On the north and south sides of the ecological island, there is about 2600 square meters of ecological purification coenosis, which can purify the lake through physical filtration, filter matrix adsorption, the decomposition and absorption of the roots of aquatic plants and other means. Through the construction of the whole system, it saves 200 million Yuan for municipal rainwater pipe network upgrade, 130 million Yuan for water saving per year, 190,000 Yuan for the remediation investment of runoff pollution, and 100 million Yuan for flood protection. Third, photovoltaic power generation system: integrated with the roof of the civic square, using immediate transmission and usage control technology for exclusive electricity and use, improving usage efficiency, saving investment and maintenance costs. The construction of the Cultural Center expresses a yearning for ecological civilization, and plays an exemplary role in the construction of eco Tianjin city.

It is hoped that the Tianjin Cultural Center becomes an exhibition center of high arts, a popularization center for culture and art, a prosperous and grand people's hall, and a civic paradise which could promote the carrier function of the city, push forwards the development of the city, and bring benefits for future generations.

大剧院夜景
Night View of Grand Theater

2

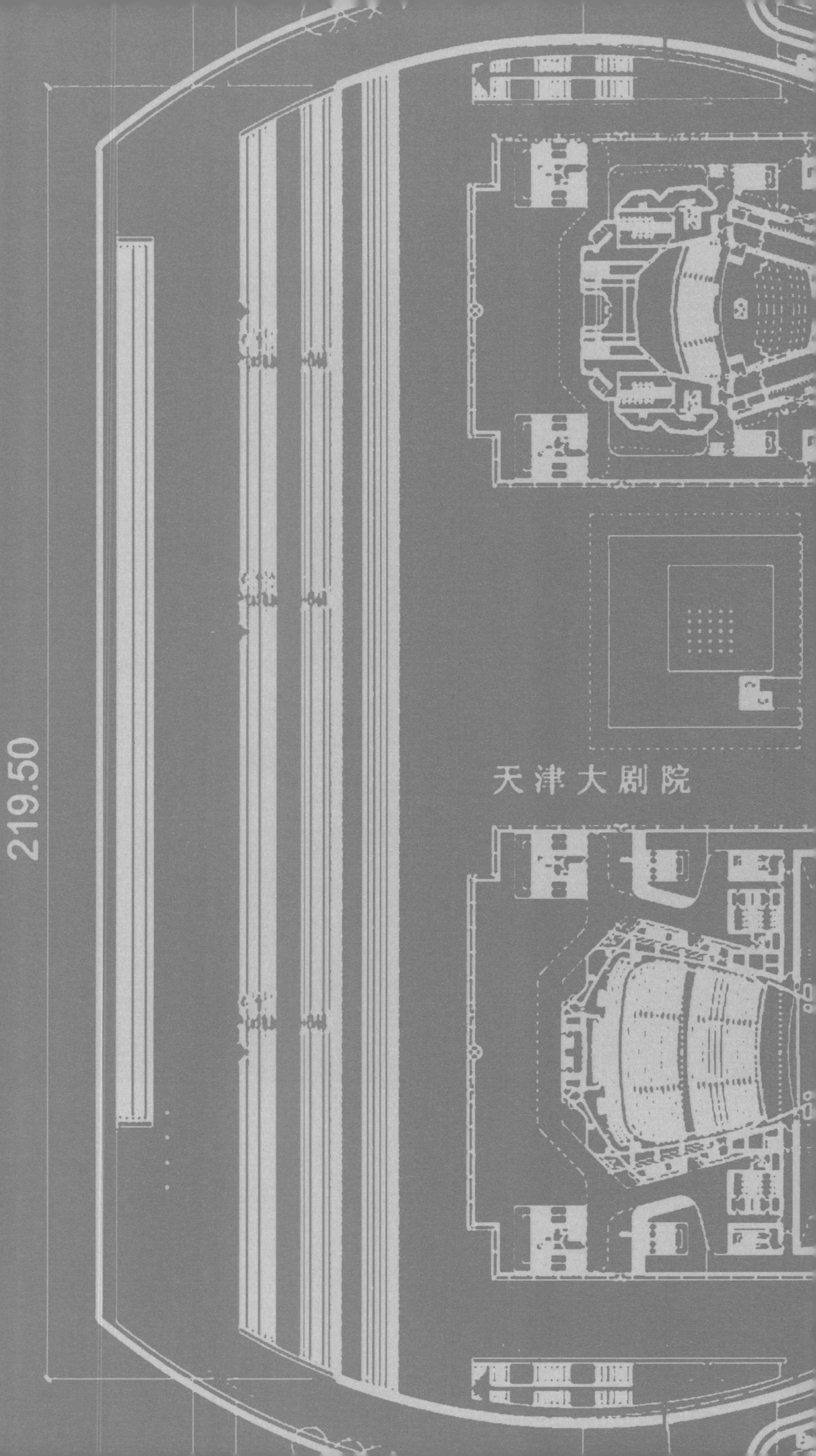

建筑设计篇

Part 2: Architectural Design

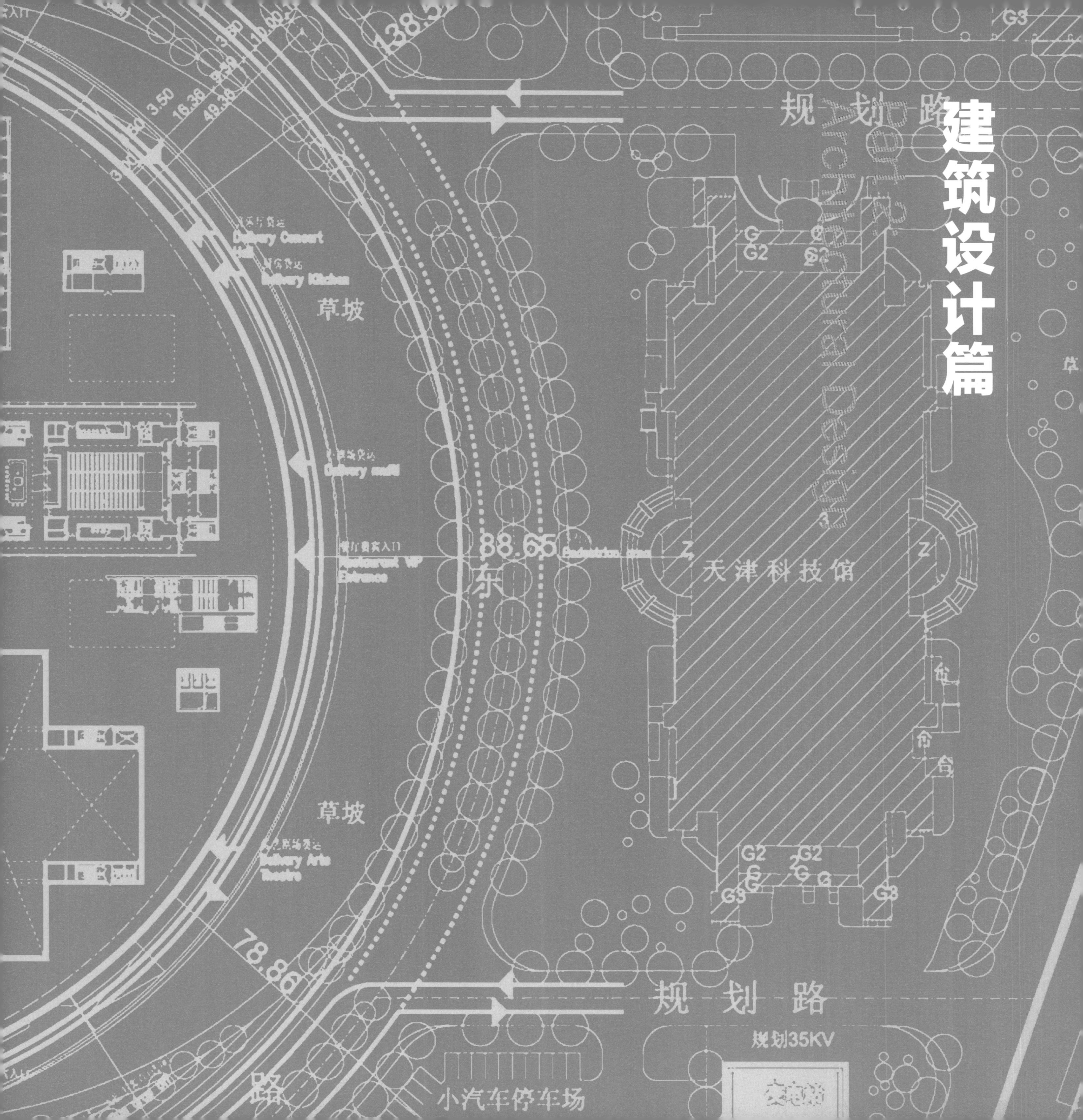

大剧院 Grand Theatre

冯-格康，玛格及合伙人建筑师事务所（gmp）
von Gerken, Marg and Partners Architects (gmp)
现代设计集团华东建筑设计研究院有限公司
East China Architectural Design & Research Institute Co., Ltd.

建筑设计

Architectural Design

摄影/Photographer：Christian Gahl

大剧院正面，临水大台阶
Front Side Waterfront Stairs

临水的大台阶作为城市舞台
Waterfront Stairs as Urban Stage

从图书馆眺望大剧院
Overlooking Grand Theatre from Library

从阳光乐园看大剧院
Overlooking Grand Theater from Sunshine Paradise

晚间大剧院的照明效果
Grand Theater at Night with Illumination

在过去的三年里，天津大剧院从规划设计到施工建成的过程中gmp与中方合作院华东建筑设计研究院有限公司一起，以前所未有的速度完成了这一设计任务。天津大剧院并非一个孤立的单体，而是天津文化中心的一个有机的组成部分。天津文化中心已于近期向市民开放，它赋予了天津崭新的面貌。建筑师最开始并不清楚将城市中心规划为文化广场这一主张源自何方，但毫无疑问，结果是振奋人心、令人赞叹的 。

狭长的水景定义了地块中央的景观公园。这一中心元素将所有建筑融为一体，新建筑和原有建筑，特别是天津博物馆、科技馆、大剧院和谐地成为一个有机的整体。

在项目的建设过程中，天津市规划局展示了专业的组织、规划能力。来自世界各地的建筑设计师能够众志成城，为同一个目标而努力，与此同时又能展现各自不同的艺术风格。可以说，天津文化广场项目是一个国际合作的优秀范例，其经验值得借鉴和推广。

至今，在建筑师、景观设计师、基础设计规划人员与城市规划及文化部门负责人举行例会、交流想法、共同对话讨论的场景仍历历在目。在这个项目中，建筑单体并不是建筑师的自说自话，而是一个建筑综合整体，是各方相互沟通呼应的成果。如亚里士多德所说：“整体大于部分之和。”

Within a possibly record-breaking planning and construction schedule of 3 years, the Grand Theatre of Tianjin was designed by gmp office and planned together with their project partner ECADI. It is not a single building, but a part of an ensemble in the central cultural park which opened in 2012. It is the new face of the city. The origin of the idea to use the city center as a place of culture is unknown to the architect. The outcome, however, is fascinating in many ways.

The elongated waterscape defines the landscape park in the center of the site. This central element integrates all the buildings, not only the new ones, but also the existing buildings as well, especially Tianjin Museum, Tianjin Technical Museum and Tianjin Grand Theater, which become a balanced unity.

The artistry demonstrated by the Urban Planning Bureau of Tianjin to realize its own master plan, the skill to enable convened architects from all over the world to find a common attitude without abandoning their own character, is exemplary and worth adopting.

Until now, the architect can clearly remember the regular meetings in which architects, landscape designers, infrastructure designers and people responsible for urban planning and culture were meeting each other, changing their opinions and making mutual dialogues. There was no architectural monologue of single buildings. The buildings constructed correspond with each other and form a unity. As Aristotle once said, "the whole is greater than the sum of its parts".

大剧院鸟瞰
Aerial View

从文化中心北侧湖岸远眺大剧院
Overlooking Grand Theatre from Northern Lake Shore of Cultural Centre

大剧院 2 层平面　Grand Theatre F2 floor plan

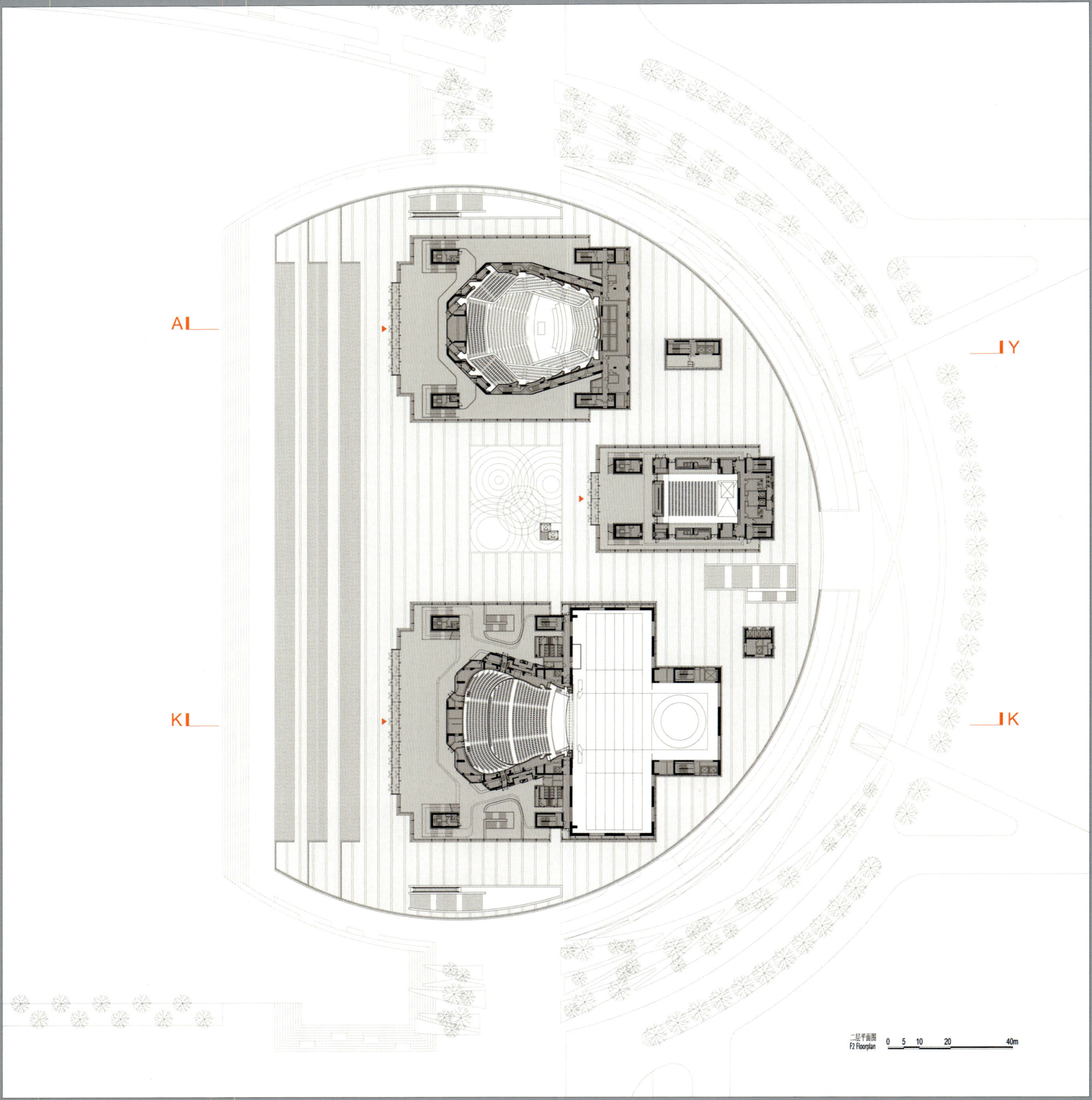

在这一项目中，户外空间的布置特别令人满意。绿地与铺面比例适中，达到平衡；湖岸的设计各具特色，景观设计精致而别具匠心。建筑师所达成的共识体现在建筑体量的比例和连贯的浅色天然石材的使用上。整个园区设计中最为值得称道的是行人流线相对于机动车流线的优先度。机动车只允许在园区外围出入。

天津大剧院的环形造型与天津博物馆相呼应，形成了文化公园东西两侧建筑间的体量对话。

大剧院设有一个5米高的平台，俯视公园，平台连接一个近300米宽的大台阶，在这里可以远眺湖景和整个文化公园，是都市生活的舞台。这里也是市民聚会、休息、娱乐的场所，美丽壮观的喷泉为文化广场增添了活泼的氛围。在宜人的夏夜，每个曾在这里欣赏音乐喷泉的人都不会忘记这一动人时刻。

大剧院的设计充分体现了gmp的设计哲学，在多样性中体现统一，于统一中实现多样性。统一的大屋顶将三座不同的剧场连接起来，是体现gmp设计哲学的一个典型范例。

剧场组织

三座剧场的室内设计与其功能相关。音乐厅完全采用木质饰板，如同小提琴的琴身。葡萄园梯田式的席位设置在中国开创了先河，突破了“鞋盒式“的常规剧院布置。演奏家们与他们的音乐成为观众注意力的焦点，声学效果极具震撼力。歌剧厅的色彩及建筑手法的运用更加大胆并富有戏剧性，与歌剧的表演内容相辅相承。线形排列的千万个光点如同苍穹的点点星光，让人联想起卡尔·弗里德里希·申克尔为歌剧”魔笛”所作的著名的舞台设计。

希望今后在天津文化中心能够举行各种高质量的文化活动，演出，以及展览。只有市民参与到文化中心的活动中来，才能使其成为属于公众的天津新地标。

In this regard, the quality of the outdoor areas consisting in their balance of landscape and pavement, the sophisticated arrangement of watersides as well as the areas with sensitive handling of important visual axes have to be highlighted with praise. The common spirit of the architects is evident in the proportion of the buildings and the consistent use of bright stone as their material. A crucial aspect cannot be stressed enough: the prioritization of pedestrian traffic over motor traffic, the latter only surrounding the peripheral of the compound.

The circular form of Tianjin Grand Theater corresponds with Tianjin Museum, creating a dialogue between these two buildings on eastern and western ends of the cultural park. A 5m high podium is rising from the park, connecting with an approximately 300m wide staircase, which overlooks the lake and the whole cultural park and becomes the stage of urban life. It is also a place for gatherings, strolling around and recreation. The beautiful and grandiose music fountains create a fantastic atmosphere of the cultural park. People who recently had the chance to enjoy the fountains and the music in a mild summer evening would never forget this striking moment.

The design of Tianjin Grand Theater fully represents the design philosophy of our office, namely, to realize unity in variety and to express variety in unity. The homogenous roof over three different halls is the best example illustrating our design philosophy.

Organisation of Theaters

The styles of the three halls relate to their contents. The music hall is cladded overall with wooden panels, which looks like the internal side of a violin case. The vineyard-like terracing introduces a new type of concert hall in China, which differs from the regular “shoebox” with performance in the front. The orchestra and the music become the focus of attention, and the intension of sound perception is almost impossible to surpass.

The color and architectural applications used in opera hall is more scenic and relates to the contents of the opera works. Thousands of linear light points look like a starry sky, which reminds of the famous stage design of Karl Friedrich Schinkel made for “The Magic Flute”.

It is hoped that various performances, exhibitions and activities with high quality will take place in cultural center of Tianjin in the future. Only when the citizens of Tianjin can participate in the cultural events, will the cultural center of Tianjin become a collective place and a new symbol of the city.

在平台上面对位于中部的多功能厅，左侧为音乐厅，右侧为歌剧厅
View of the multifunctional hall with the concert hall on the left and the opera theater on the right

大剧院的屋顶悬浮在升起的平台层之上，位于作为城市生活舞台的大平台上可以将湖面以及文化中心尽收眼底

The roof is floating on the rising podium. On this podium of an urban stage, one has a wonderful overview on the lake and cultural centre

三个主建筑体块间的空隙，创造了东西向的视觉通廊
The gaps among the three building blocks create a visual passage from east to west

通过大屋顶的中空部分，自然光照可以到达平台层，使平台层上的空间感受更加通透明亮
Through the void of the roof, natural lights can reach the podium and make the space on podium more tranparent and brighter.

音乐厅
Concert Hall

侧入口及通往 2 层平台的台阶
Side entrance and staircase to the podium level

大剧院的屋顶悬浮在升起的平台层之上，位于作为城市生活舞台的大平台上可以将湖面以及文化中心尽收眼底
The roof is floating on the rising podium. On this podium of an urban stage, one has a wonderful overview on the lake and cultural centre

歌剧厅前厅
Foyer of the opera Hall

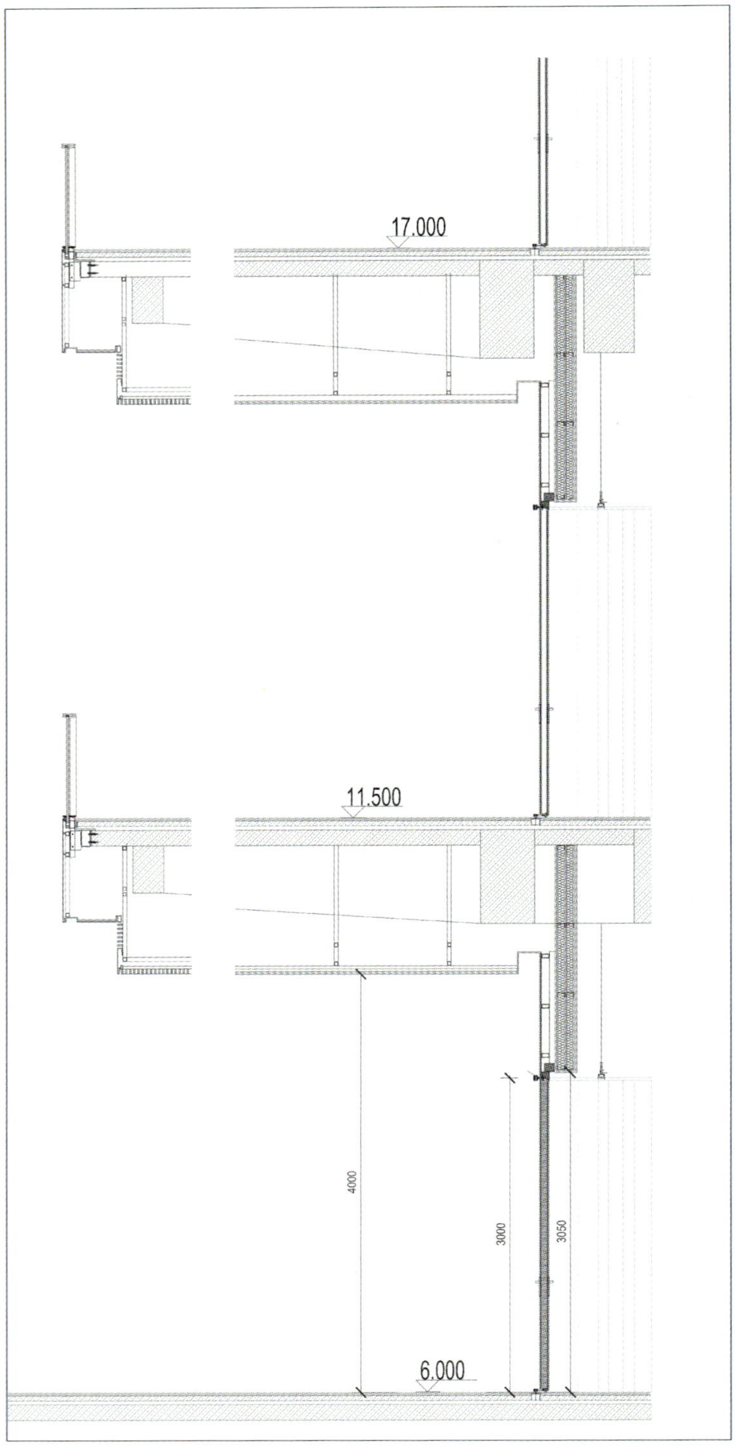

前厅回廊剖面细部图
Section Details of Foyer Gallery

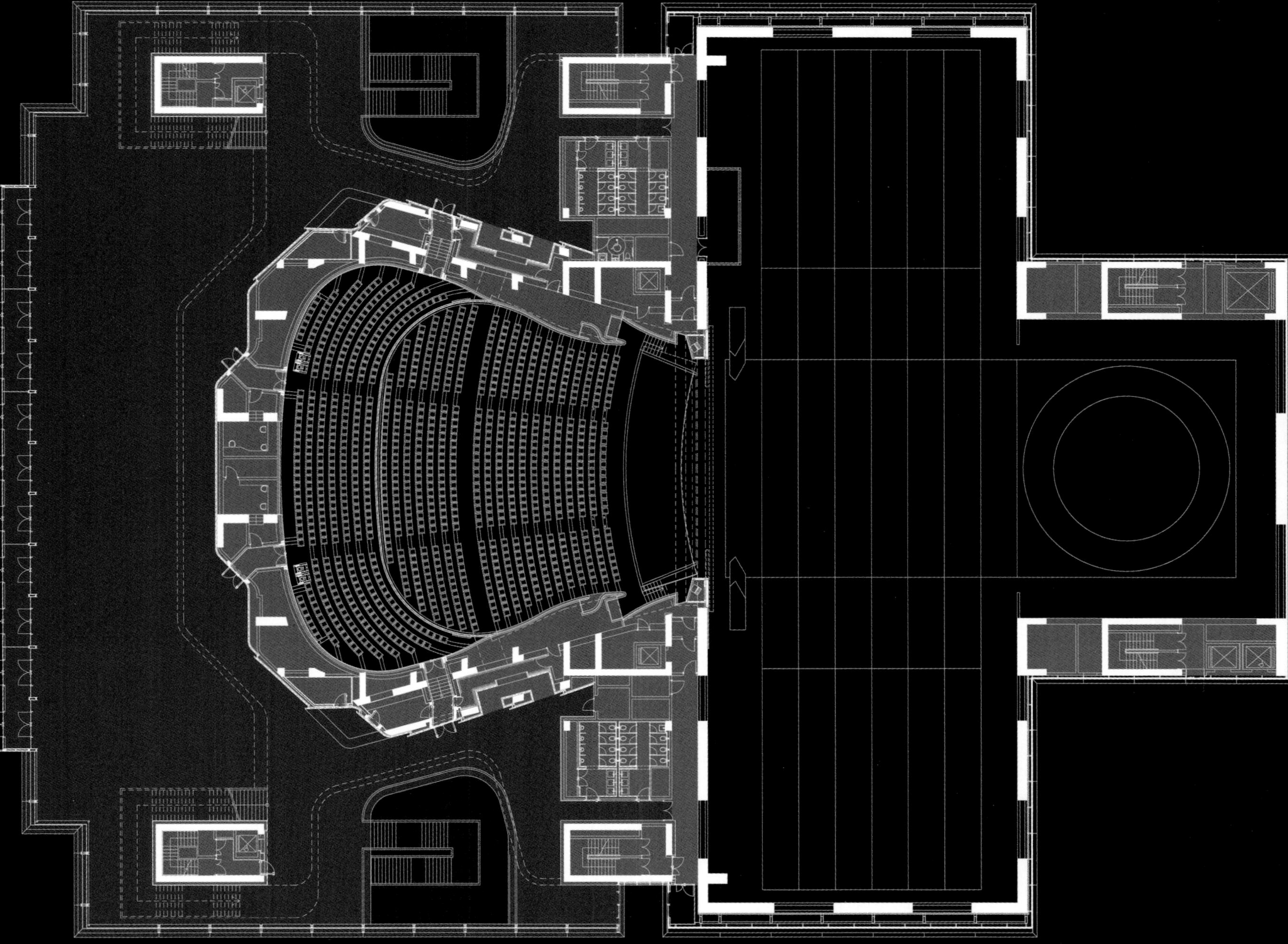

歌剧厅 2 层平面
Opera Hall F2 Floor Plan

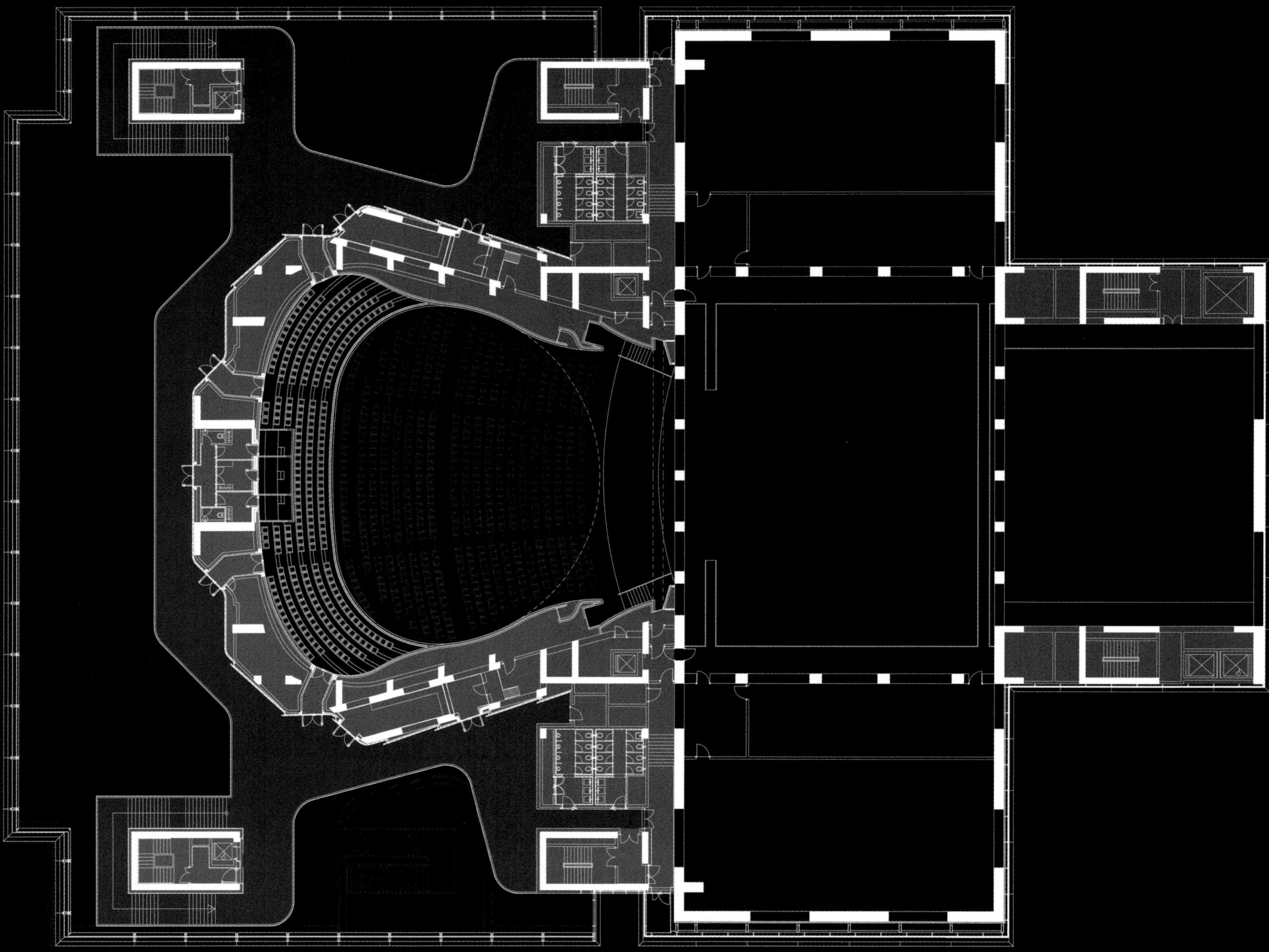

歌剧厅 3 层平面
Opera Hall F3 Floor Plan

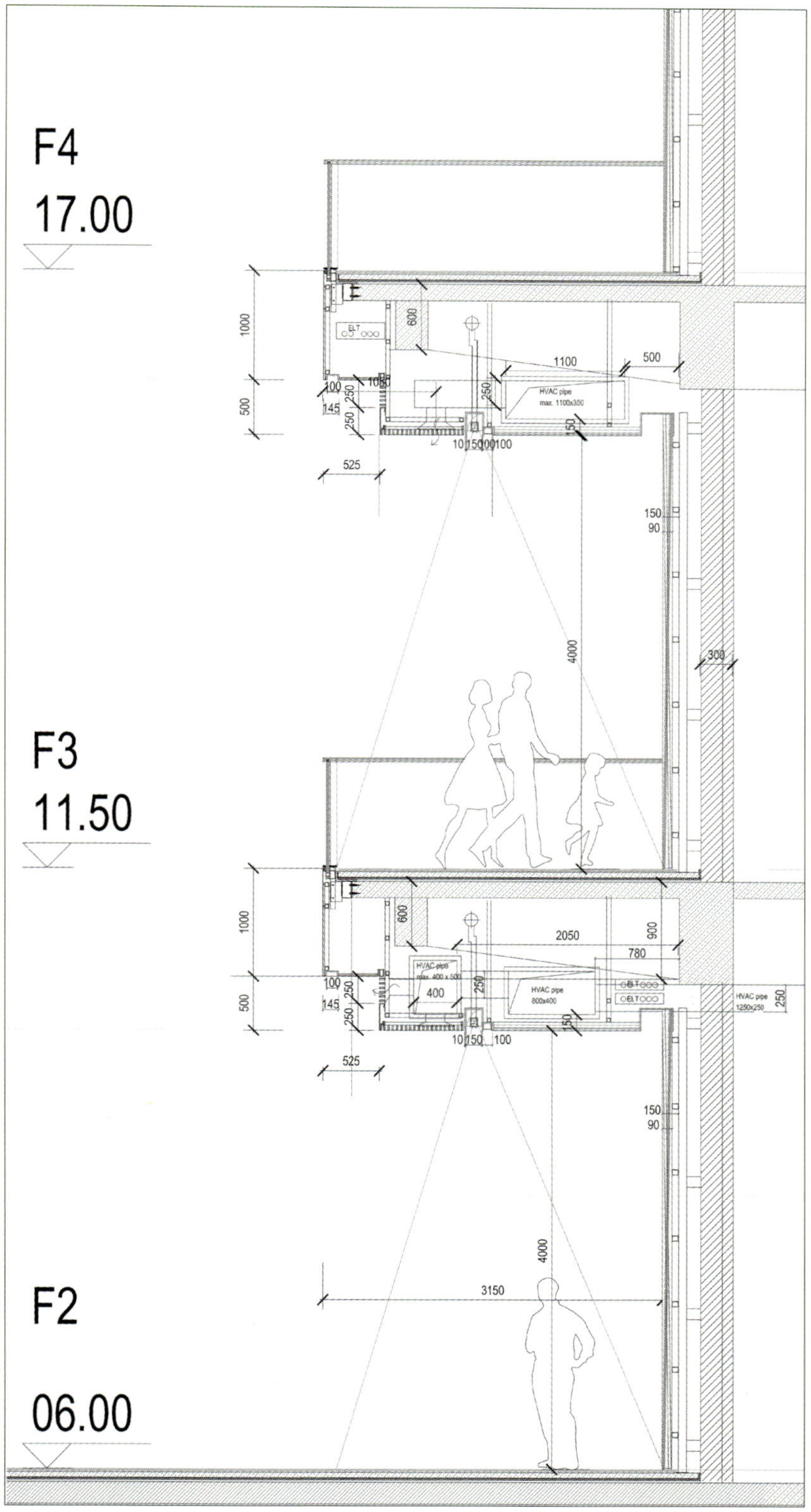

前厅回廊剖面细部图
Section Details of Foyer Gallery

歌剧厅前厅
Foyer of the Opera Hall

歌剧厅一层前厅 Opera Hall F1 Foyer

歌剧厅一层前厅存衣处　Opera Hall F1 Foyer Cloak room

歌剧厅为穹顶的造型，平面规整简洁对称，空间尺度大，是十分理想的大型演出场地。观众席由池座和三层楼座组成。厅内的水平向百叶墙面不但是建筑外观的一种延续，同时更强调了流畅的空间造型曲线。厅内的灯光设计效果更是烘托出了这个1600座的大剧场的恢宏气势。

The Opera Hall with a dome has a regular layout. Its large-scaled space is symmetric and concise and is an ideal space for huge performances. The auditorium consists of stall and balconies of 3 floors. The wall with horizontal lamella is not only an extension of architectural appearance, but also underlines the fluent spatial curve. The lighting design furthermore strengthens the grandeur of this theatre with 1600 seats.

图片来源 : ullstein bild - AKG

大剧院观众厅
Grand Theater Auditorium

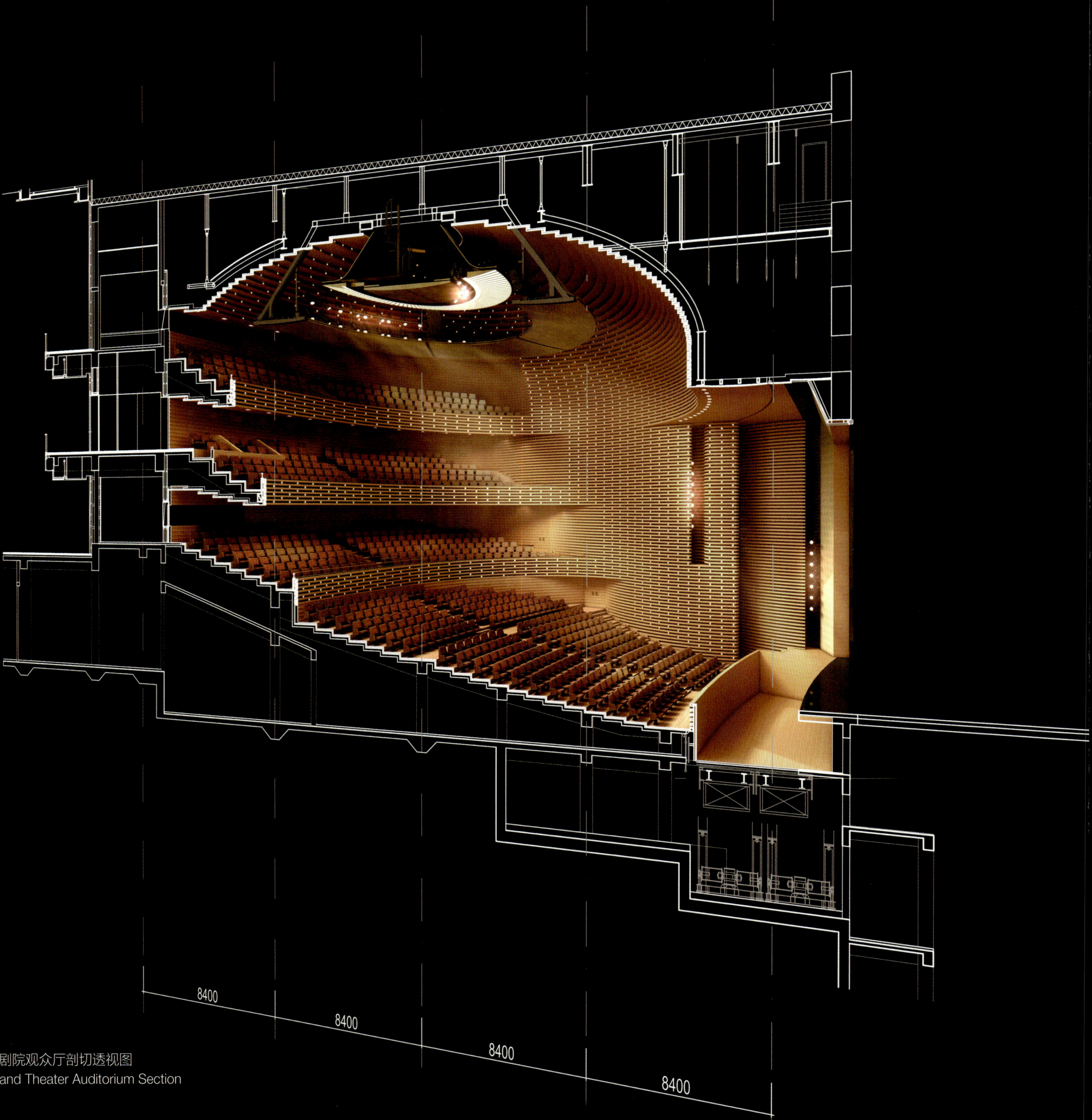

大剧院观众厅剖切透视图
Grand Theater Auditorium Section

大剧院观众厅 Grand Theater Auditorium

大剧院观众厅，从舞台朝观众厅方向观看
Grand Theater Auditorium, Looking from Stage to Auditorium

大剧院观众厅，从楼座朝大幕方向观看
Grand Theater Auditorium, Looking from Balcony Towards the Stage

大剧院观众厅，楼座栏杆细部
Grand Theater Auditorium, Railing Details in Balcony

大剧院观众厅，楼座及反声罩
Grand Theater Auditorium, Balcony and Anti-Sound Cover

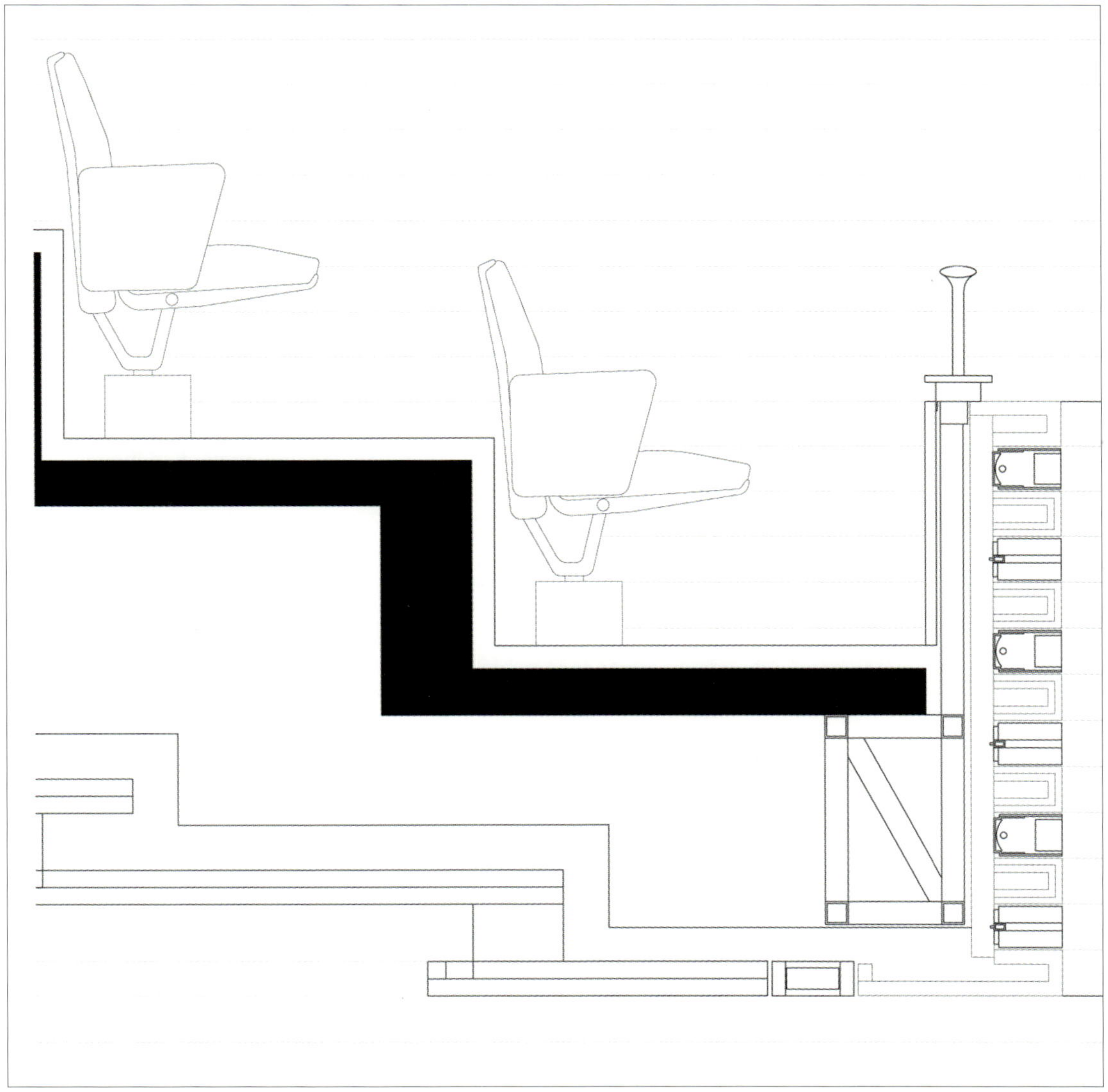

音乐厅前厅
Music Hall Foyer

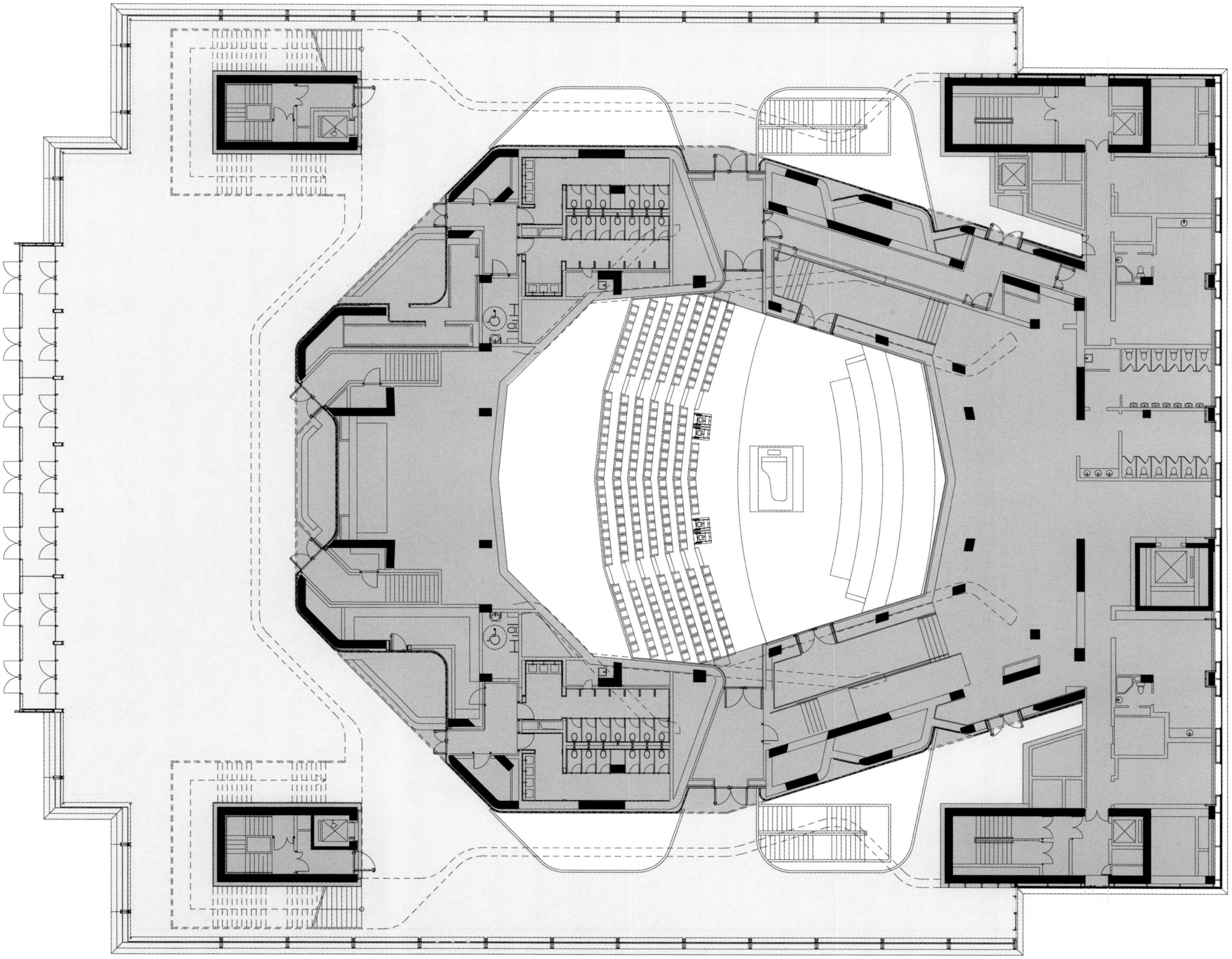

音乐厅 2 层平面
Music Hall F2 Floor Plan

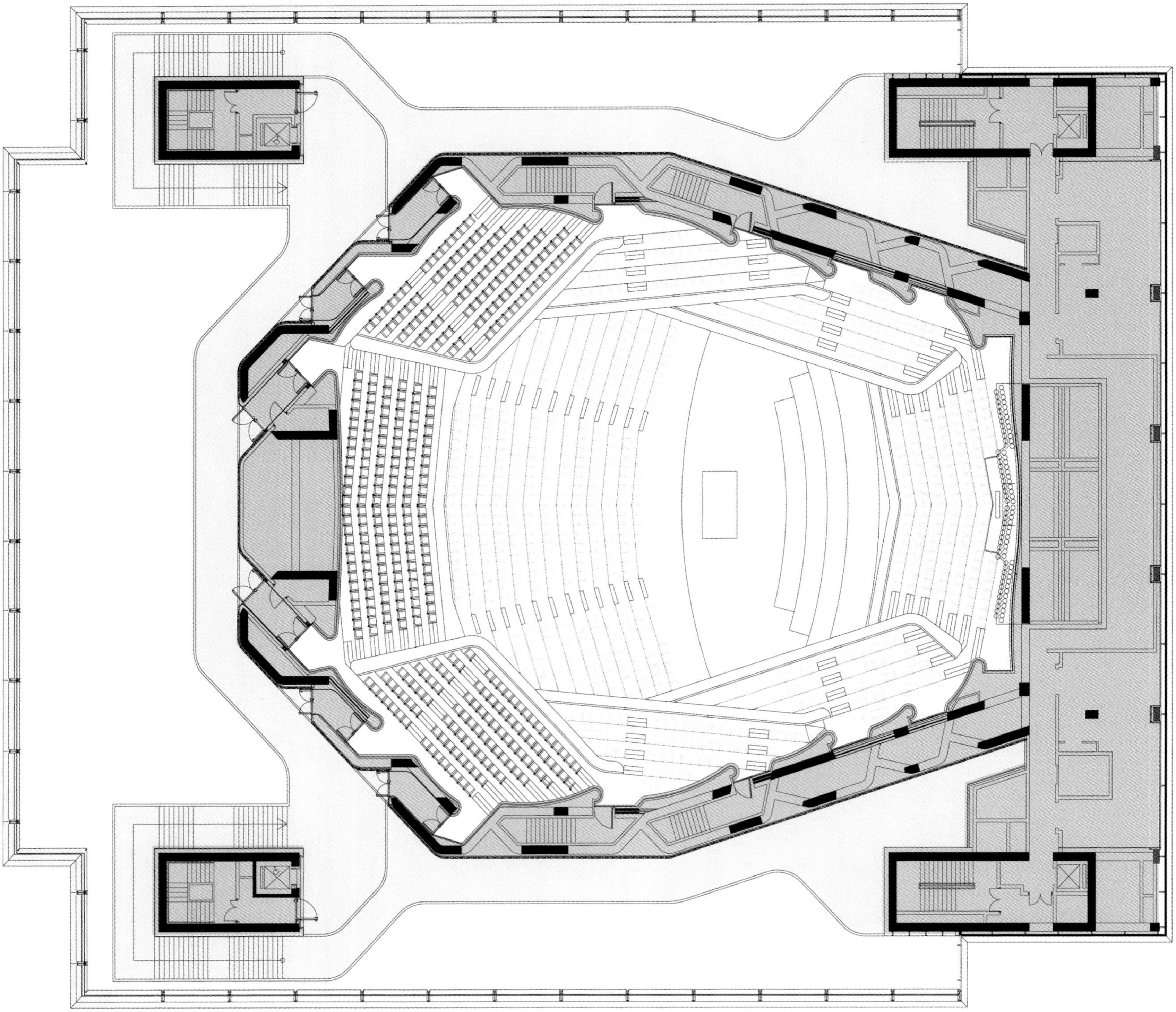

音乐厅 4 层平面
Music Hall F4 Floor Plan

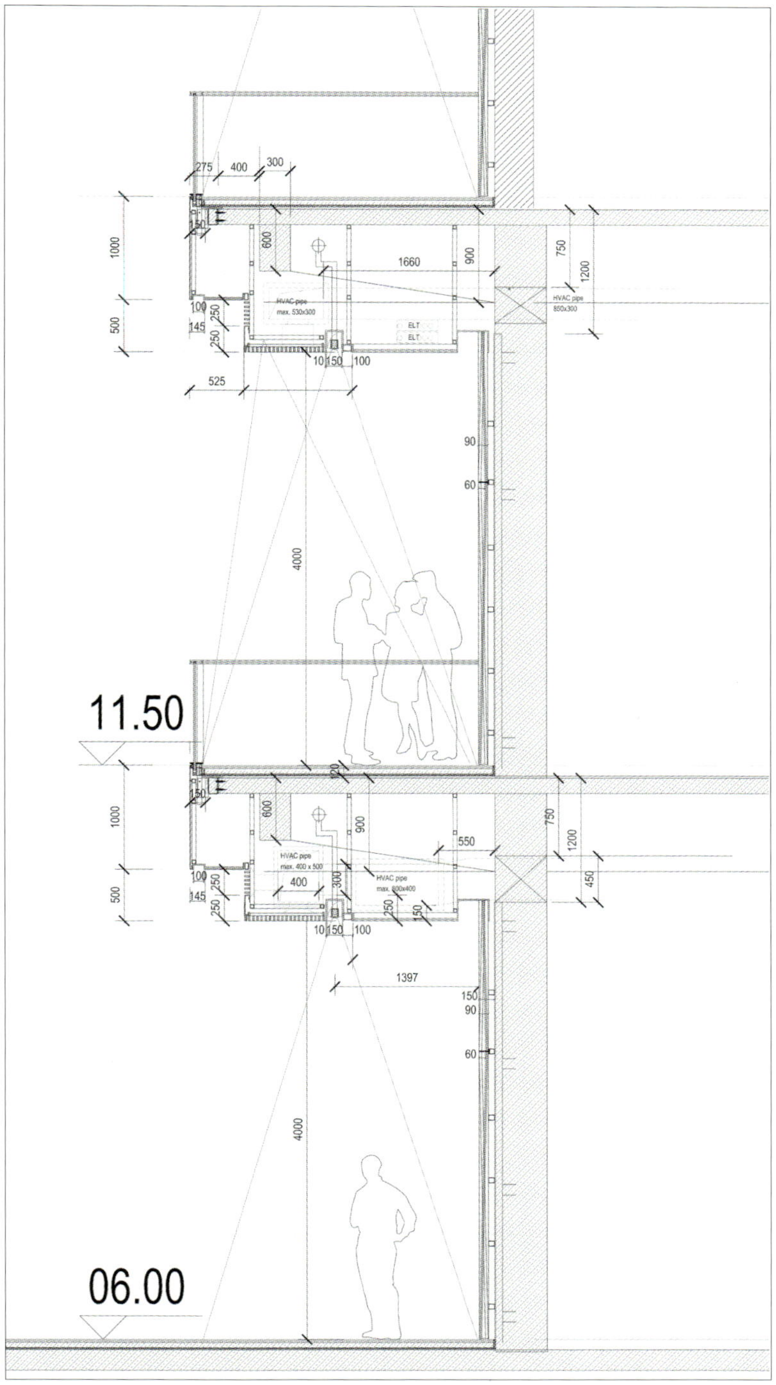

前厅回廊剖面细部图
Section Details of Foyer Gallery

音乐厅前厅，楼梯通往 1 层及 2 层前厅
Music Hall Foyer Stairs Leading to F1 and F2 Foyer

图片来源 : ©Angelo Cavalli/Robert Harding World Imagery/Corbis

图片来源 : ©Image Source/Corbis

音乐厅具有1200座的规模，其观众厅采用葡萄园式的叠台式布局，给观众提供了各个角度的观演可能。室内的墙面同样以横向百叶为主要元素，通过几组不同的圆形，形成波纹交错的室内装饰效果。

The Music Hall has 1200 seats. Its auditorium has a replicate layout like vineyard, so that the audience can enjoy performances from every angle. Like the Arts Theatre, the wall of Music Hall also has horizontal lamellas as main element. By using several groups of circles, an interior wavy effect is created.

音乐厅观众厅，从楼座朝舞台方向观看
Music Hall Auditorium, Looking from Balcony to Stage

音乐厅观众厅，从池座朝舞台方向观看
Music Hall Auditorium, Looking from Stall towards Stage

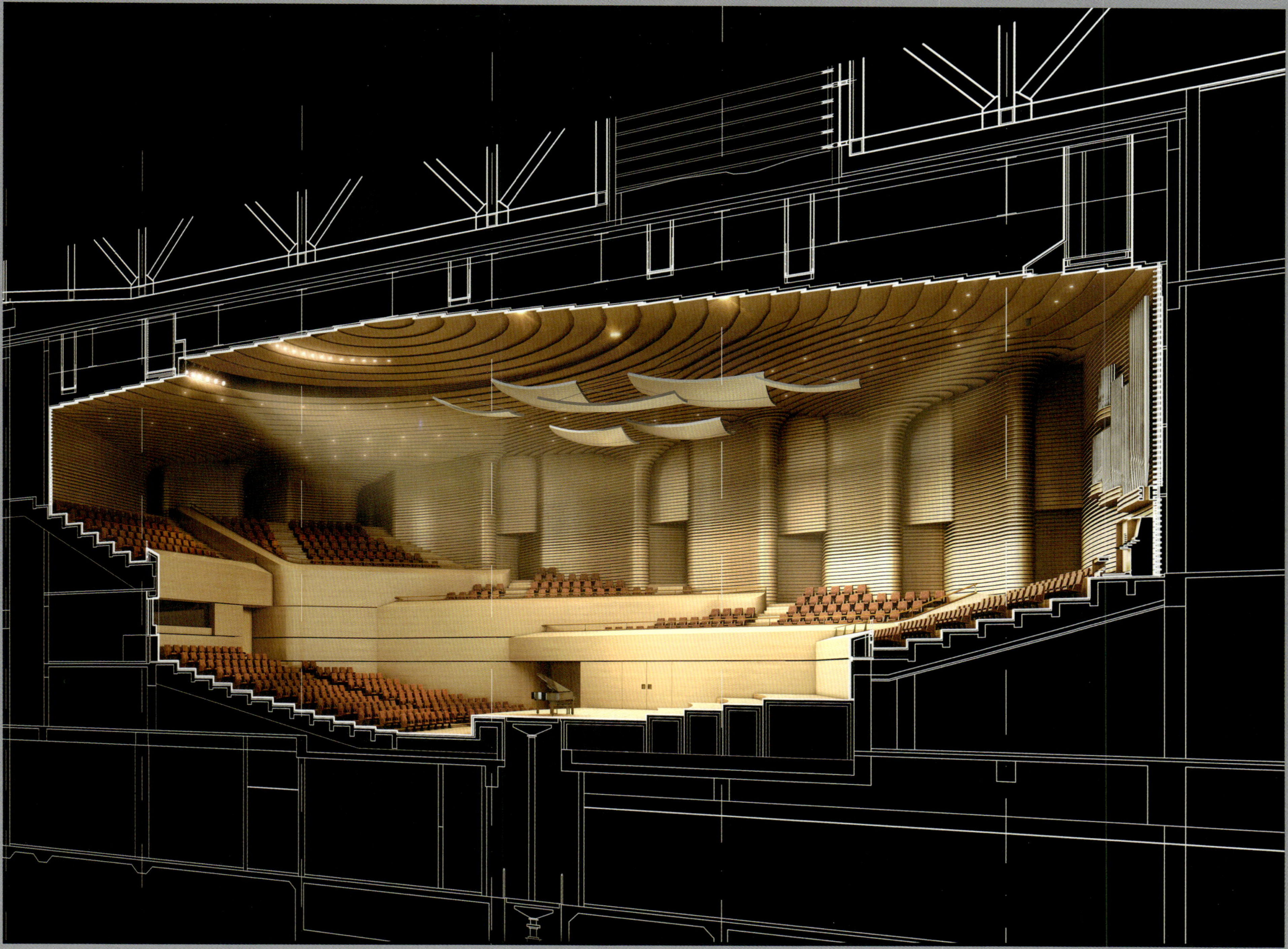

音乐厅观众厅剖切透视图

Section details of Music Hall Auditorium

音乐厅观众厅，从楼座朝舞台方向观看
Music Hall Auditorium, Looking from Balcony towards Stage

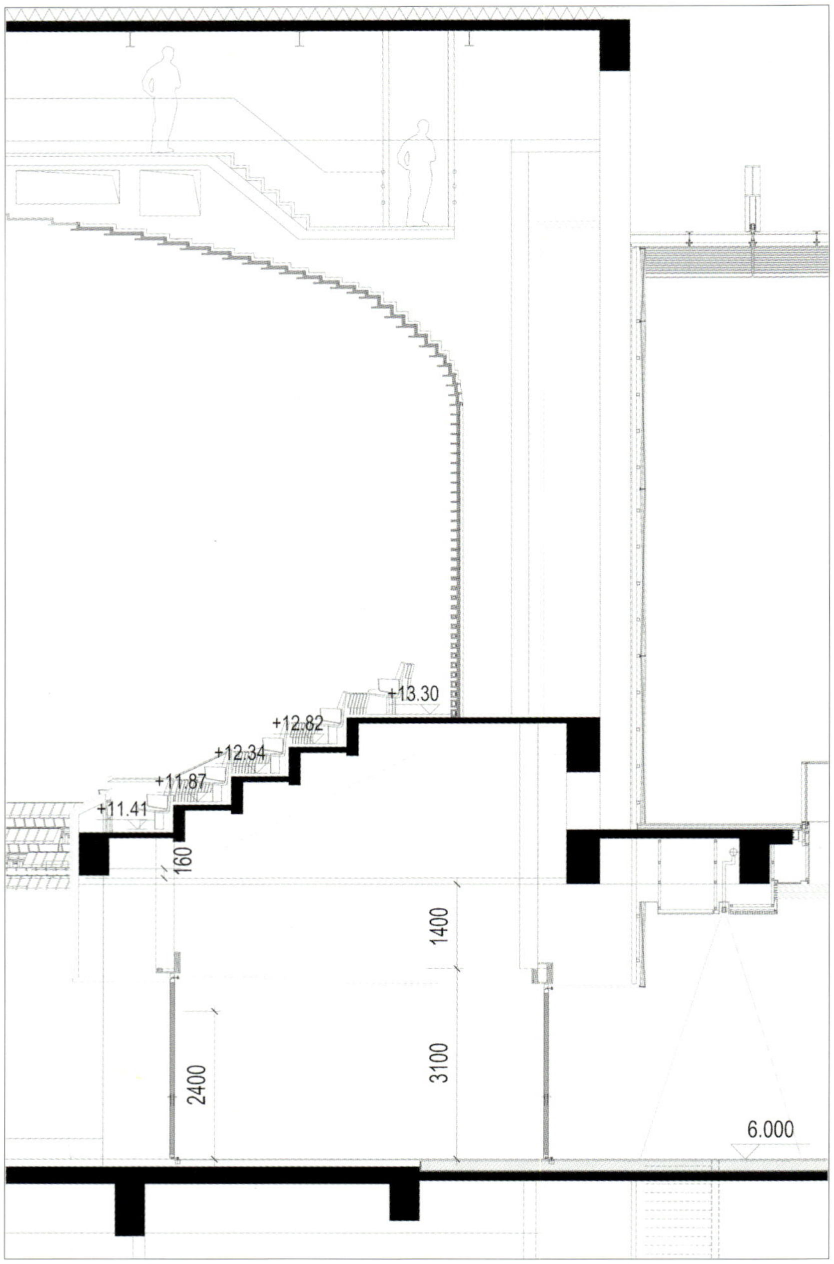

音乐厅观众厅，楼座及细部
Music Hall Auditorium Balcony and Its Details

音乐厅观众厅楼座
Music Hall Auditorium Balcony

小剧场外观
Exterior appearance of Multi-functional Hall

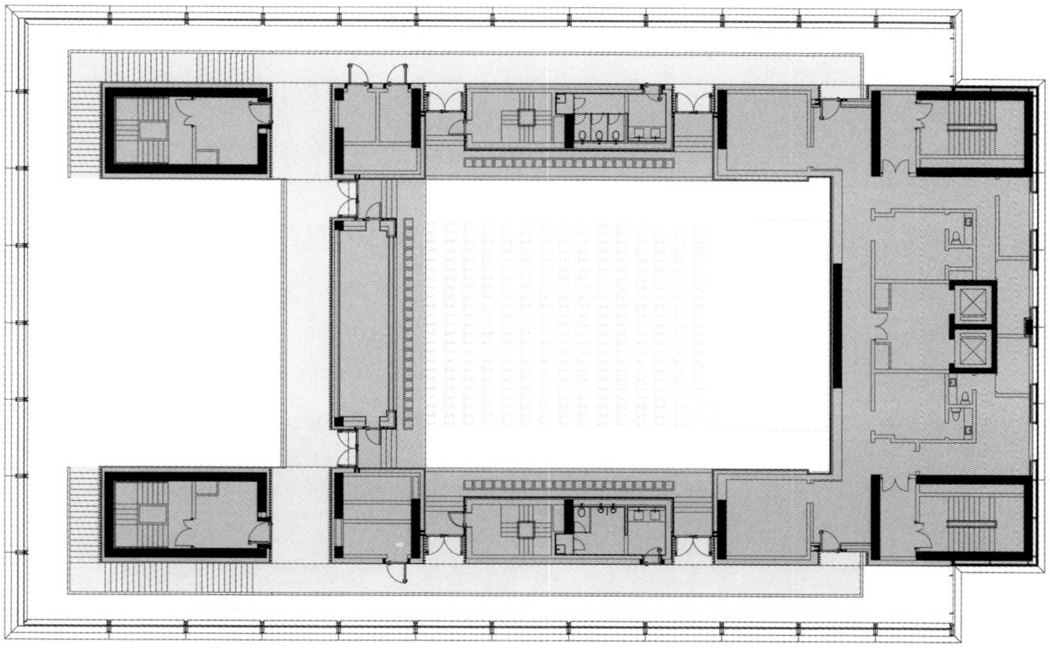

小剧场 3 层平面
Multi-functional Hall F3 Floor Plan

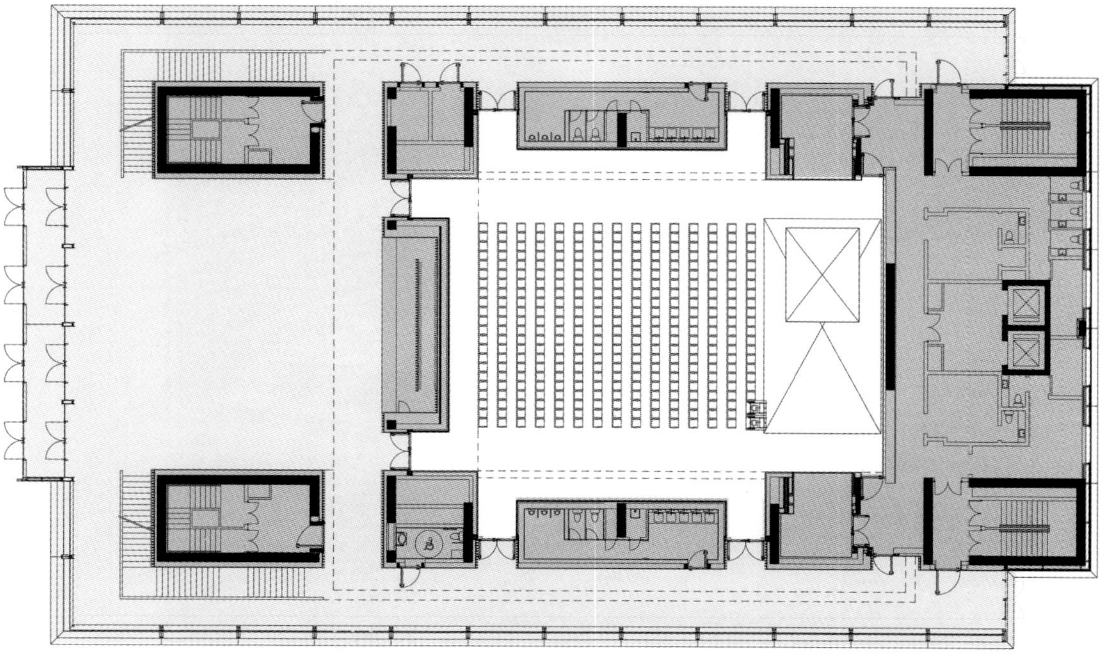

小剧场 2 层平面
Multi-functional Hall F2 Floor Plan

小剧场则十分突出它的技术性美感，墙面采用深色的声学镶板，顶部的造型则主要考虑了舞台机械的需求。内部空间为十分规整的长方形，舞台和观众席通过台下机械可以排列成不同的形式来满足不同的演出要求。

In Multi-functional Hall the technical aesthetics is underlined. The wall uses dark acoustic panels. The roof formation mainly considers the demands of stage techniques. The interior space is a regular rectangle. Through mechanical facilities under stage, stage and auditorium can be arranged as different forms in order to satisfy different requirements on performance.

小剧场观众厅
Multi-functional Hall Auditorium

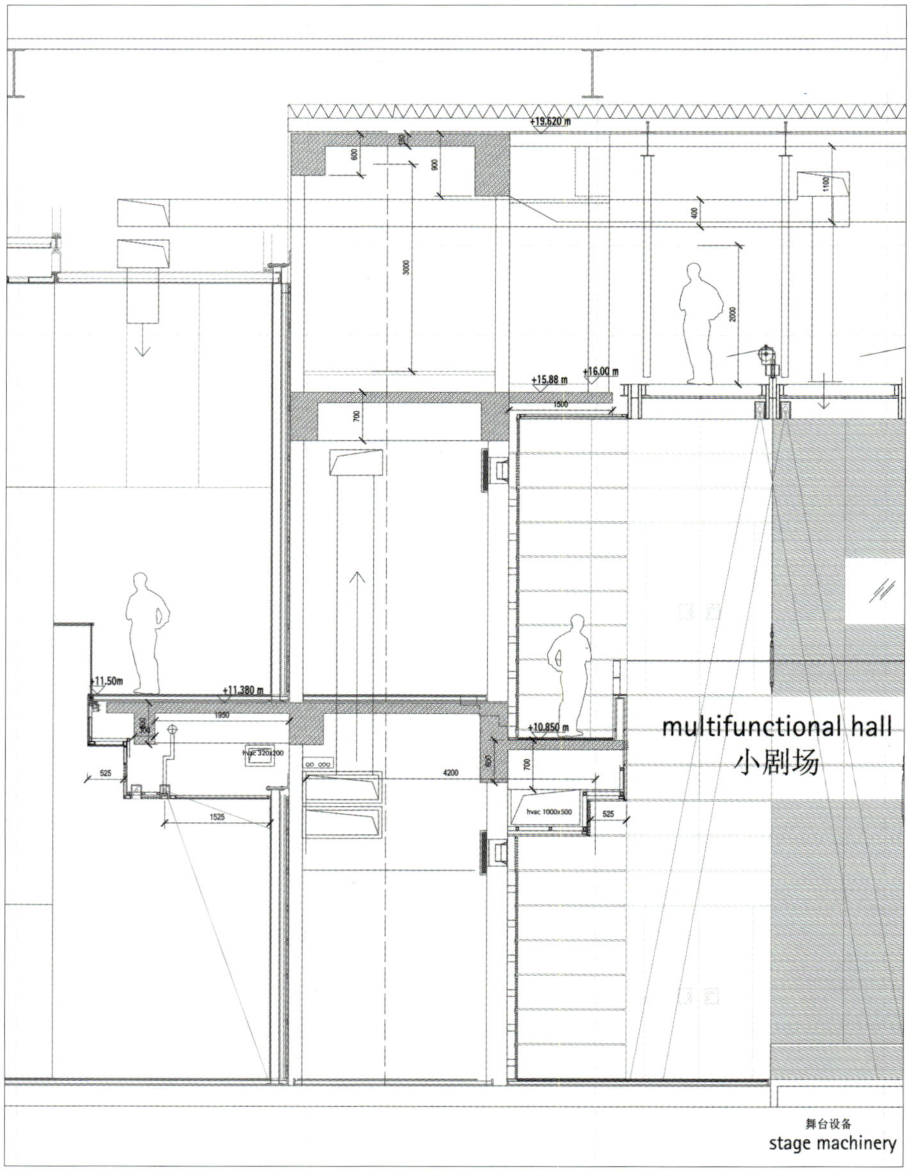

小剧场前厅及观众厅剖面细部图
Section Details of Multi-functional Hall Foyer and Auditorium

小剧场前厅 Multi-functional Hall Foyer

2 层平台层入口及主立面
F2 Main Entrance and Main Facade

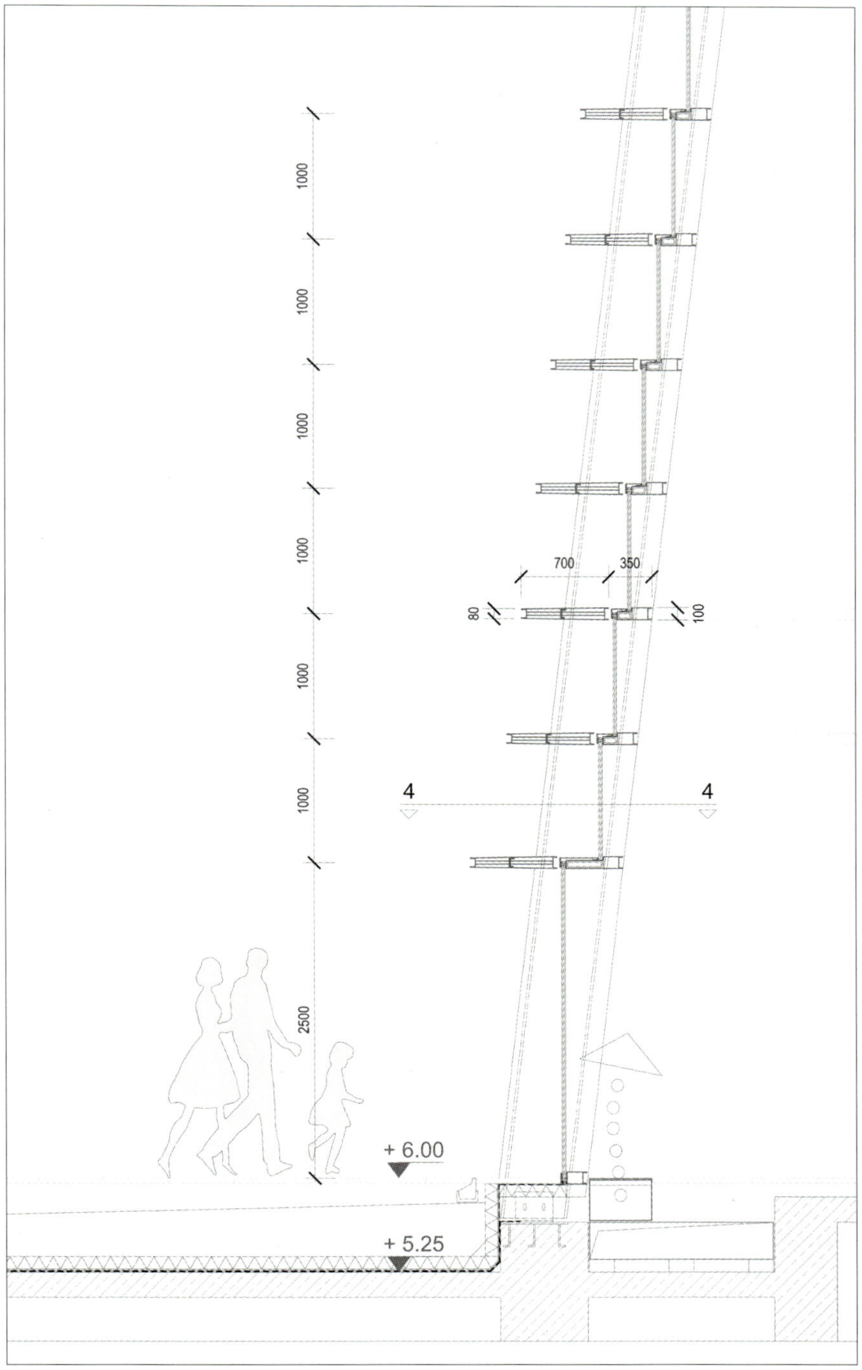

主立面幕墙细部图
Details of Main Facade

从前厅透过玻璃幕墙远眺湖景
Looking over Lakescape from Foyer Through Glass Facade

大屋顶幕墙剖立面细部图
section details of roof facade

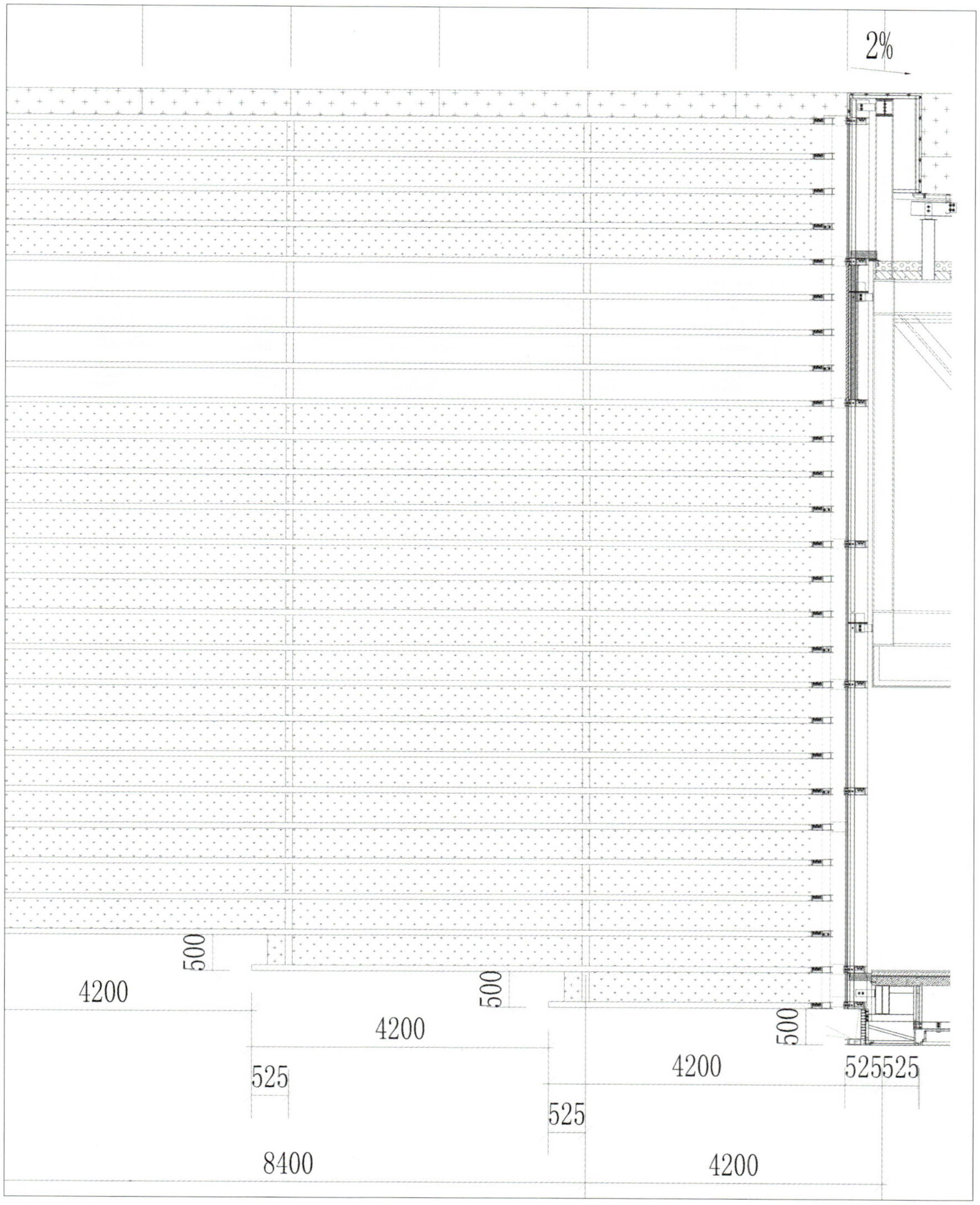

在 2 层平台上，从音乐厅朝歌剧厅方向观看
Looking from Music Hall towards Opera Hall on F2 Podium

位于东侧的出租车通道
Taxi Approach on the Eastern Side

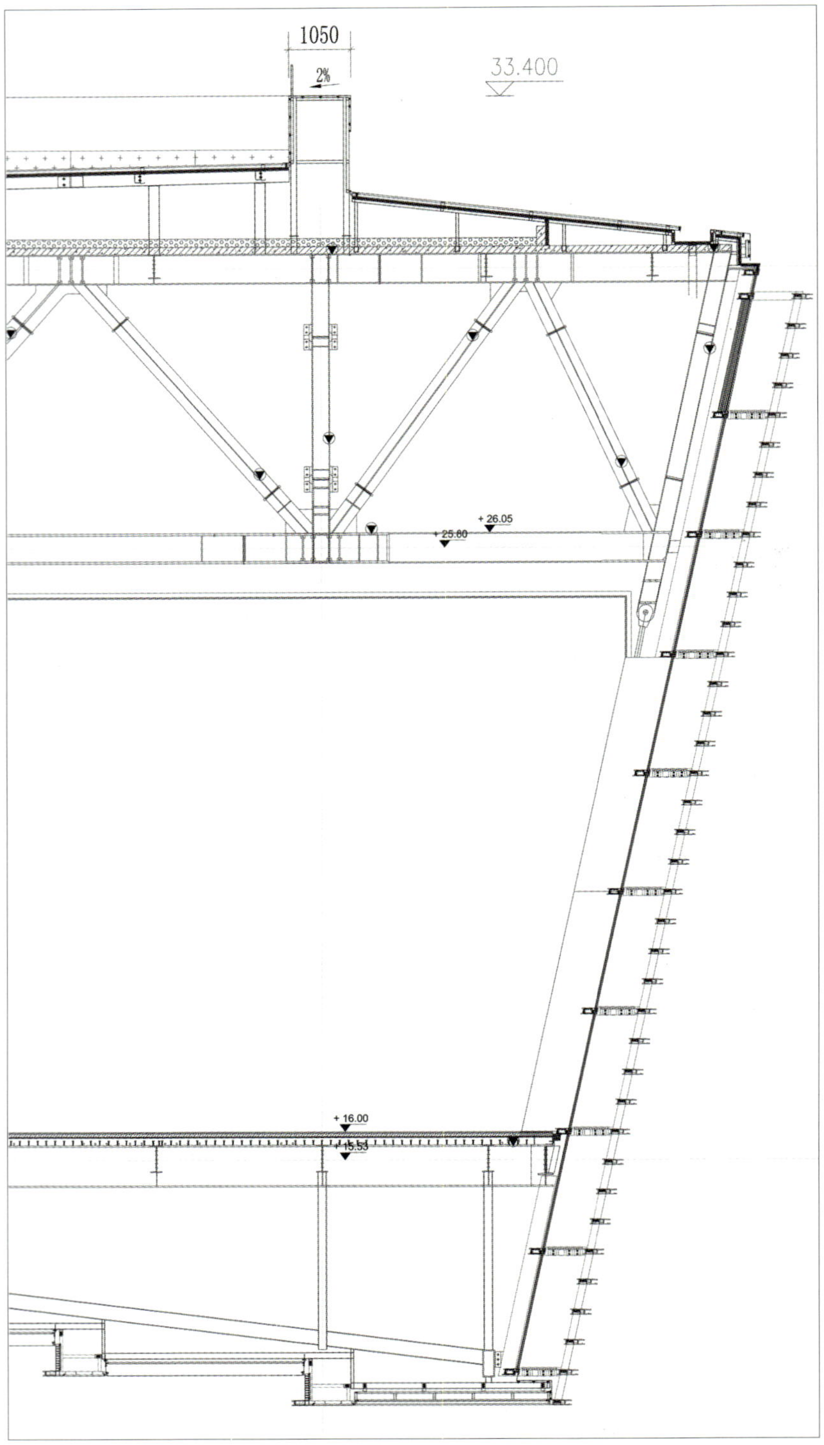

大屋顶幕墙剖面细部图
Section Details of Roof Facade

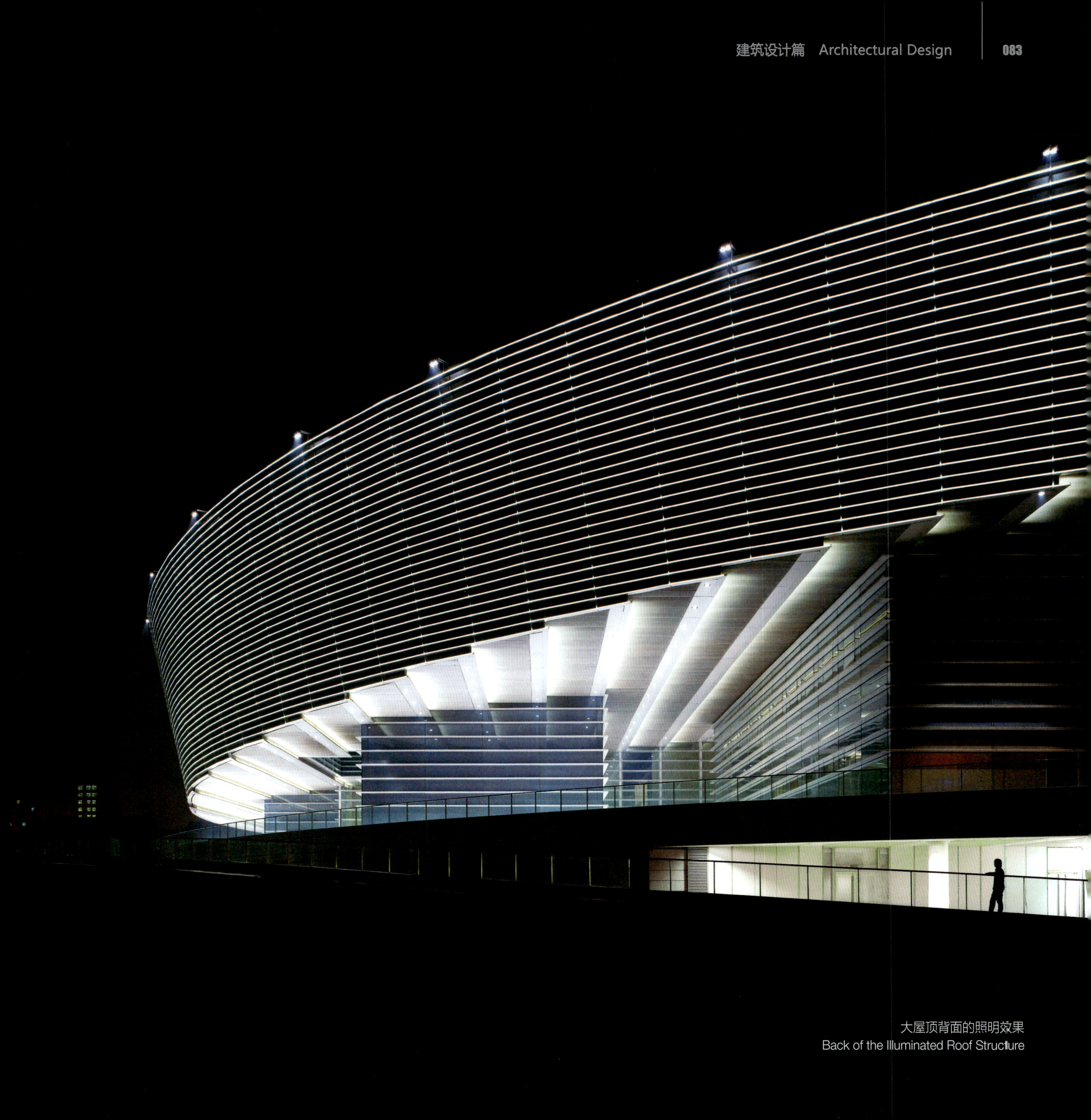

大屋顶背面的照明效果
Back of the Illuminated Roof Structure

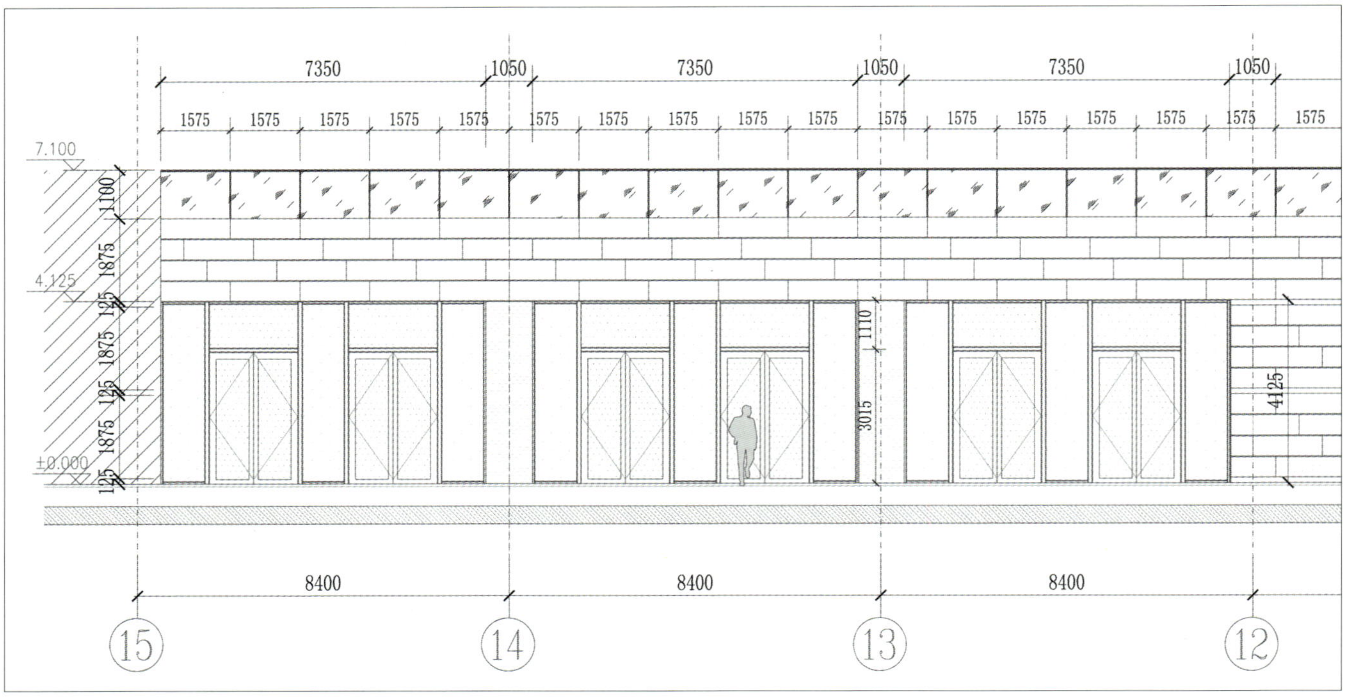

石质基座，侧入口立面
Stone Base, Side Entrance Perspective

侧入口及通往 2 层平台的台阶
Side Entrance and Staircase to the Podium Level

大屋顶钢结构

Steel Structure of the Roof

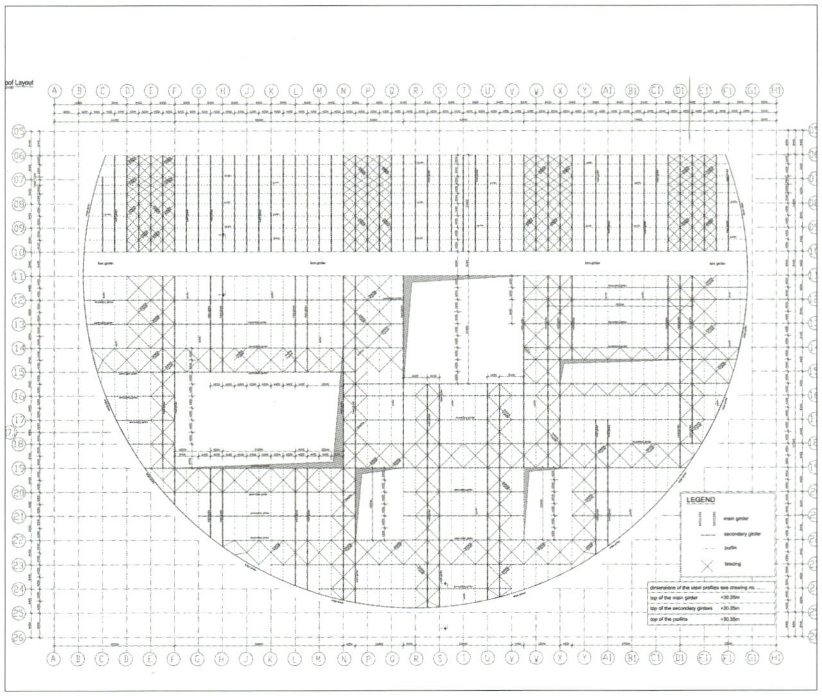

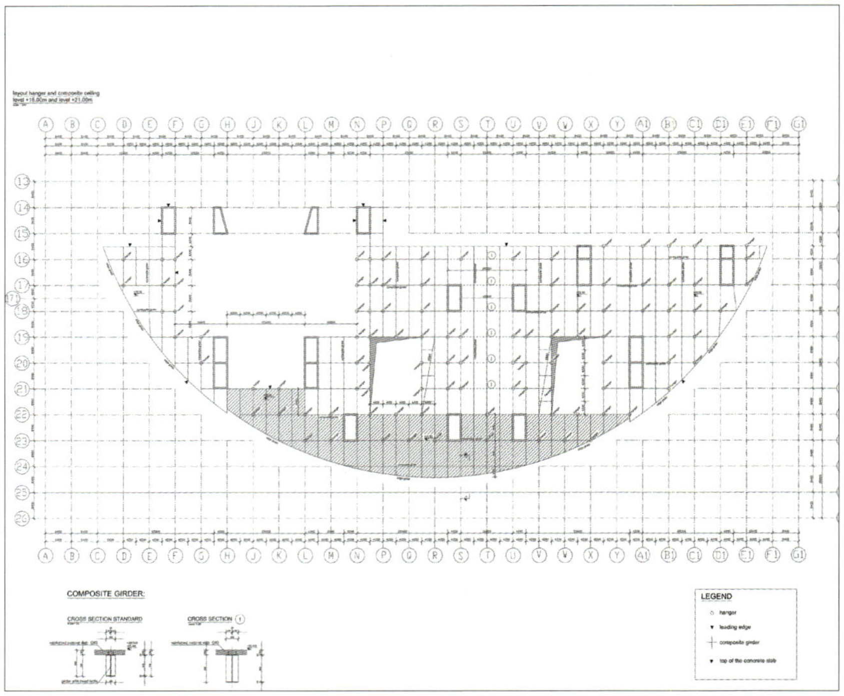

大屋顶结构模型

Structure Model of the Roof

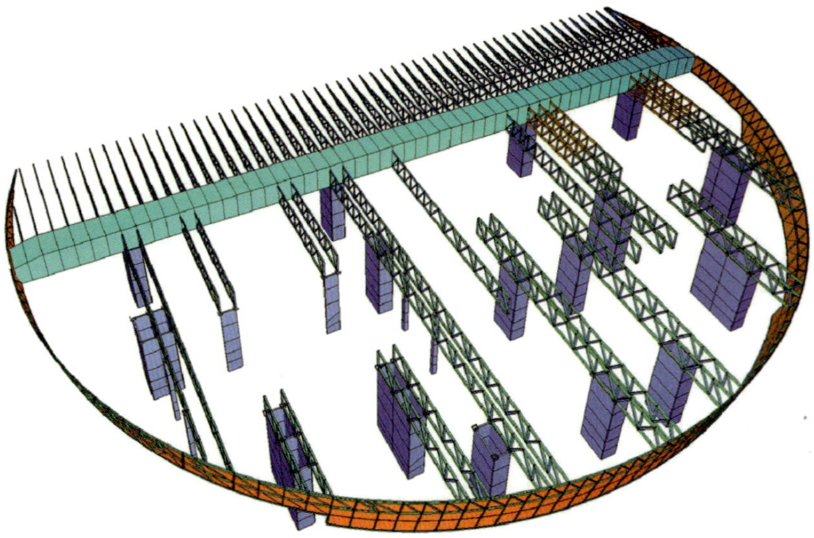

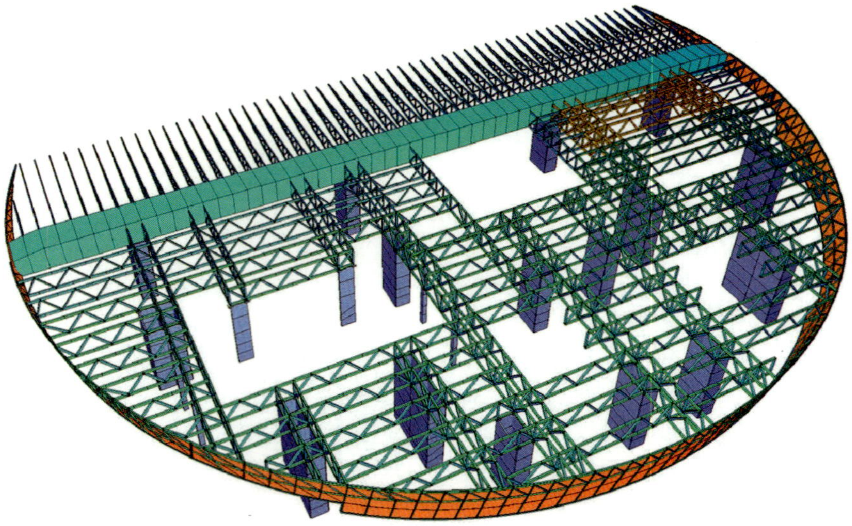

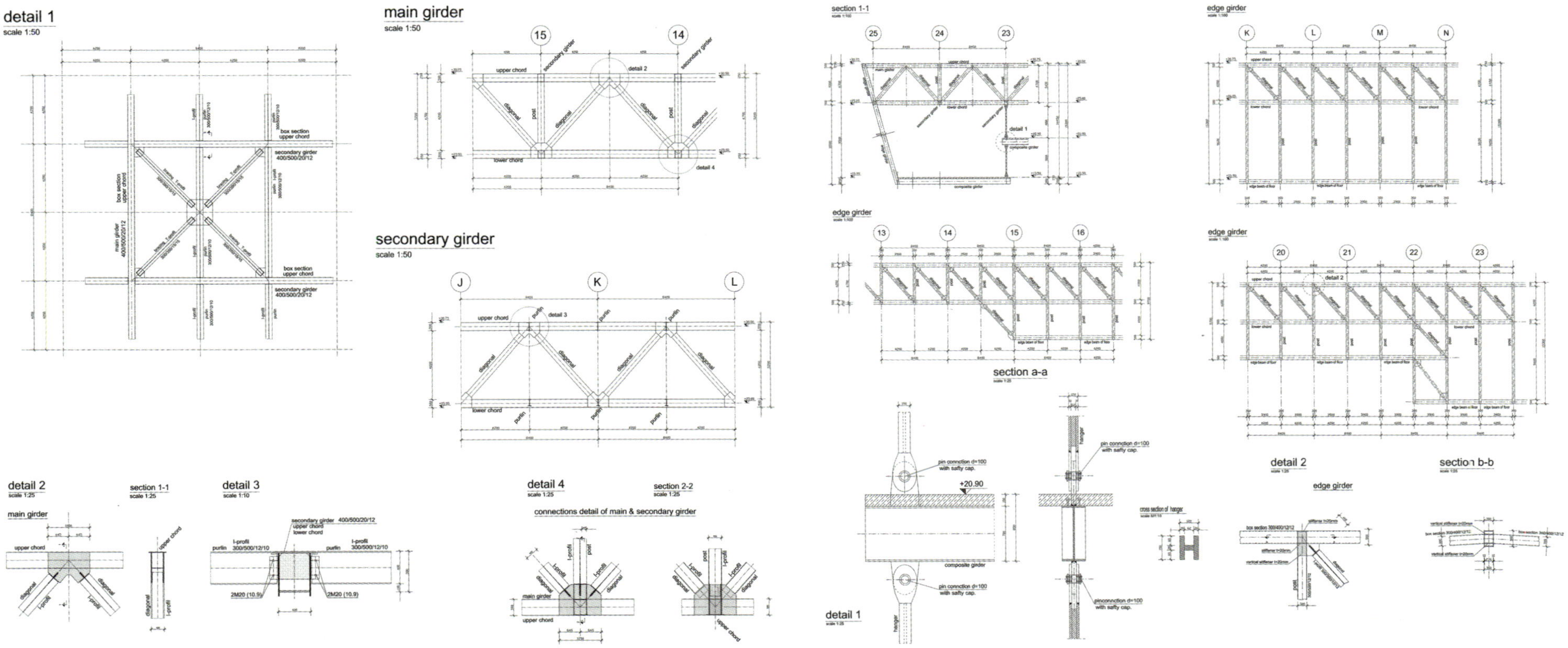
detail 1
scale 1:50
box section
upper chord
secondary girder
400/500/20/12
main girder
400/500/20/12
main girder
scale 1:50
upper chord
secondary girder
detail 2
diagonal
post
lower chord
detail 4
secondary girder
scale 1:50
upper chord
detail 3
purlin
diagonal
lower chord
section 1-1
edge girder
section a-a
detail 1
composite girder
pin connection d=100
with safty cap.
+20.90
hanger
cross section of hanger
detail 2
section b-b
detail 2
scale 1:25
section 1-1
scale 1:25
detail 3
scale 1:10
upper chord
I-profil
300/500/12/10
purlin
2M20 (10.9)
detail 4
scale 1:25
section 2-2
scale 1:25
connections detail of main & secondary girder
main girder
upper chord
post
diagonal
I-profil

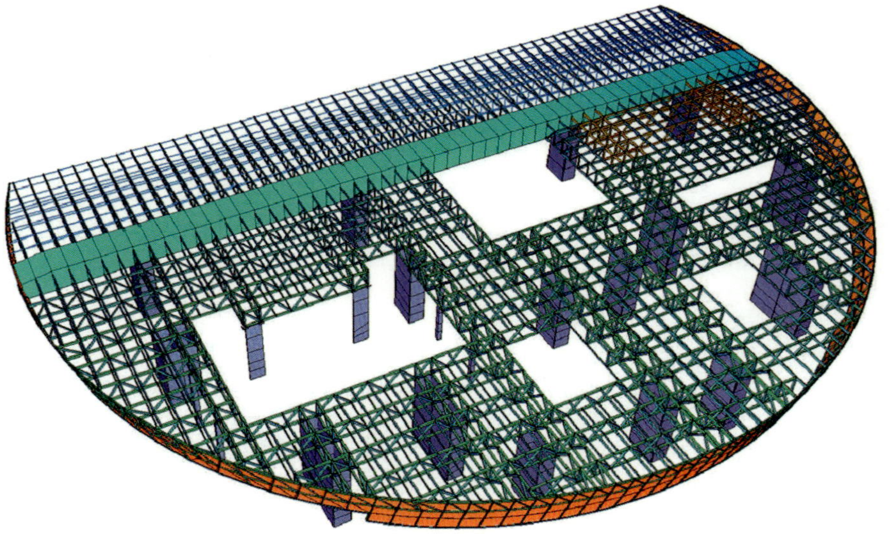

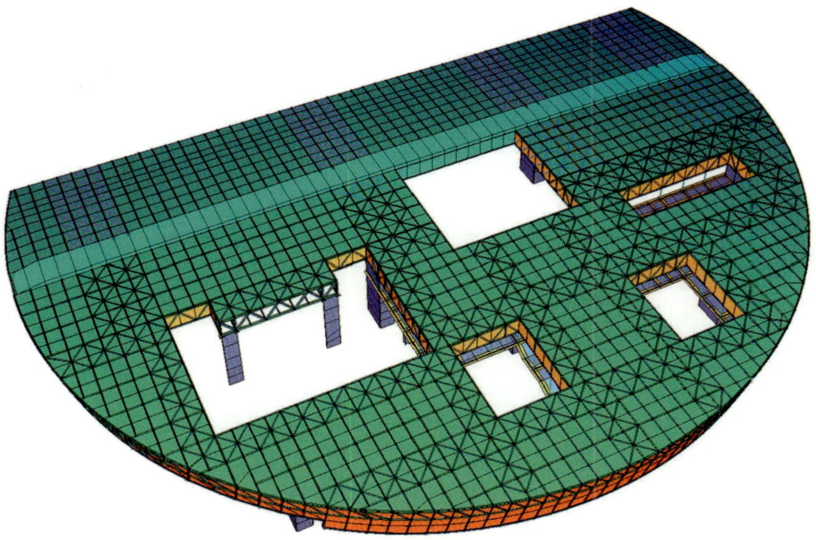

结构设计

本工程的上部结构体系是利用建筑垂直交通以及机电设备用房布置成钢筋混凝土筒体作为支撑大型钢结构屋盖的竖向构件，使之成为本工程的主结构体系；其次，利用三个剧场的周边建筑墙体设置钢筋混凝土剪力墙或框架柱作为支撑剧场本身钢屋盖的竖向结构，使之成为本工程的次结构体系。上述主次结构体系通过平面交通形成的局部楼板相互连接。

主结构体系为本工程的设计重点。由于本工程钢结构屋盖需覆盖三个剧场（三个次结构），其面积巨大（长向235m，短向160m），除地震及风荷载作用外，温度作用对屋盖体系本身以及混凝土支撑结构的影响上升为主要问题。因此，利用中部筒体与钢屋盖采用固定铰支座的连接方式，将主要的地震与风荷载作用传递至中部筒体；周边筒体与钢屋盖的连接采用单向、双向滑动支座，以释放钢屋盖产生的温度效应；从而达到改善整体结构抗震性能与减小温度应力的统一。

另外钢屋盖前部、公共活动区域上方建筑师要求开一个42米×37米的大洞口，造成该部位33.6米大悬挑钢桁架根部无法贯通。采用常规的空间钢桁架结构时，计算分析结果表明，悬挑桁架端部的变形远超过规范限值，无法满足正常使用极限状态的要求。为了解决上述问题，创新性地提出采用在桁架四周外覆钢板的空间桁架结构。

分析结果表明，设置外覆钢板后形成的蒙皮抗扭桁架能够大幅减小悬挑钢桁架根部的扭转角，从而有效控制大悬挑桁架的端部变形。在此基础上进一步研究了蒙皮的应力状况，对其作屈曲稳定分析，通过设置波纹钢板解决了蒙皮的屈曲稳定与现场焊接难题。另外与传统的实腹式箱形梁相比，箱形抗扭桁架的蒙皮板很薄（仅12mm左右），能够大幅降低用钢量，节约结构造价。

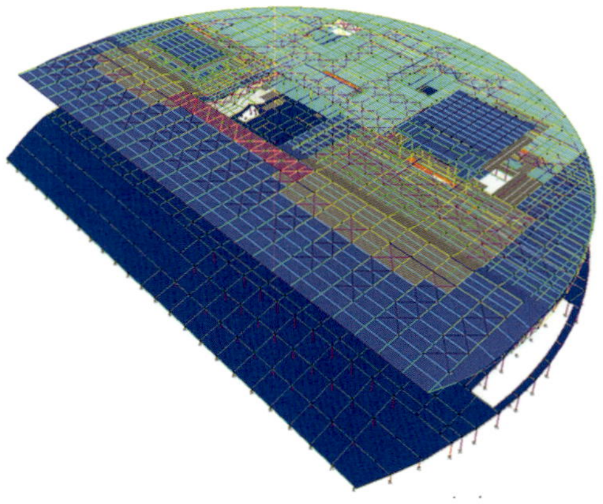

结构模型
Structure Model

悬挑桁架
Cantilever Truss System

Structural Design

The primary structural system of the project is made up of reinforced concrete tubes, composed of the vertical transportation and electromechanical equipment rooms, supporting a large steel structure roof. And reinforced concrete walls or frame columns of three theatres constitute the secondary structural system, supporting each steel structure roof. Both systems are connected through the floor.

The design of the primary structural system is a key process. Due to the steel structure roof covers three theaters with a huge area of 235m × 160m, besides the earthquake and wind actions, the temperature turns out to be a main influence. Therefore, immovable hinge supports are used to connect the central tube with the steel roof to transfer both the earthquake and wind actions to the central tube; while peripheral tubes are connected to steel roofs through unidirectional or bidirectional movable supports in order to release the temperature effect. In such a way, not only the seismic performance can be improved, but also the temperature effect can be reduced.

Besides, a hole with an area of 42m × 37m is set in front of the steel roof, just above the public region, which makes a cantilever steel truss of 33.6m in length cannot get through. If a traditional steel truss is adopted, the deflection of its end will greatly exceed the code limited value. However, an innovative scheme of wrapping steel plates around the spatial truss has been proposed.

It was found that the torsion angle at the fixed end of a cantilever steel truss can be greatly reduced when wrapped with steel plates, so that the deflection of the truss end can be effectively controlled. Besides, the stress state of the steel plate skin was further studied. The buckling analysis showed that buckling stability of the skin can be solved using corrugated steel. Moreover, compared with the traditional solid-web box girder, a box truss with a 12mm-thikck skin would greatly reduce steel consumption.

暖通空调设计

空气调节系统

剧场观众厅、舞台、音乐厅、小剧场、前厅、排练厅等采用全空气定风量空调系统。

剧场观众厅、音乐厅空调系统按观众区前后、上下分区域设置。采用座椅送风，中上部回风的空调送回风方式。送风口结合座椅设置，处理后的空气经由座椅送风口送到每个座位，形成自地面向上的气流流型。座椅送风温差控制在3.7~4℃范围内。

前厅采用不同送风温度的分层空调方式，为达到较为满意的室内热环境，沿幕墙设下部送风口，上送和侧送风口采用温控型风口，大厅下部回风，冬季采用低温热水地板辐射采暖系统作补充。

演出技术用房采用变制冷剂流量多联分体式空调系统。

行政办公用房、演出用房（化妆间、演员休息等）采用风机盘管+新风空调系统。

通风系统设计

观众厅、舞台、音乐厅、小剧场、排练厅均设置机械排风系统。综合剧场观众厅、 音乐厅排风系统设置热回收装置。

演员化妆休息室设置机械排风系统。

耳光室、追光灯室、调光柜室、灯控室、声控室、机械控制室等设置机械通风系统。

防排烟系统设计

防烟楼梯间、合用前室、消防电梯前室设机械加压送风设施。

综合剧场、音乐厅、观众入口大堂等依照建筑消防安全性能化设计方案设置消防防烟系统、排烟设施

消声与隔振

根据观演建筑声学要求进行消声和减振设计。

空调冷源和热源由区域能源中心提供。

安装过程中的空调设备
AC during Installation

座椅送风口
Ventilation under Seat

HVAC Design

Air conditioning system

Constant-volume air conditioning system is adopted at theater auditorium, stage, odeum, little drama theater, lobby and rehearsal hall.

The systems of theater auditorium and odeum are set separately according to different orientation, which applies chair-back supply air mode, returning air from the mid and upper area of the room. Delivery tuyere installed at each chair back feeds air flowing from bottom to top. The temperature difference of supply area is controlled within 3.7-4℃.

Stratified air conditioning mode with different supply air temperatures at different areas is used at lobby to achieve a better indoor thermal environment with low-temperature hot water radiant floor heating system as supplementary in winter. Several air outlets are designed to put at the lower part of curtain wall. Besides that, other outlets are installed at the upper and side area adjusted by temperature.

Performance studios chose variable refrigerant volume multi-separate air conditioning system.

Offices, dressing rooms and lounges utilize fan coil plus fresh air system.

Ventilation system

Theater auditorium, stage, odeum, little drama theater, rehearsal hall, dressing rooms and lounges boast mechanical ventilation systems. In addition, the ventilation systems of concert hall and auditorium have heat recovery equipment.

Mechanical ventilation systems are set at ear-light room, spot light room, adjusting light room, light controlling room, sound controlling room and etc as well.

Smoke control system

Smoke proof stairwell and emergency lift and smoke free shared-atria are designed with forced supply air system.

Comprehensive theater, odeum, waiting hall are set up exhaust systems and smoke prevention systems on the basis of relative construction safety performance and design requirements.

Muffler and vibration isolation

The design of noise elimination and vibration control is following requirements and demands of architectural acoustics.

Cold and heat sources of HVAC are supported by regional energy center.

给排水设计

给排水专业设计有室内的给水、排水、雨水、消防等系统，以及室外消防总体设计。

一层及地下室用水均直接利用市政自来水压力，二层及以上用水采用水箱和变频调速水泵联合供水方式。厕所冲洗用水采用中水。由于剧场建筑的演出特点决定了其用水的集中性和间断性，热水系统采用了分散式和集中式结合的供应方式：在公共卫生间洗脸盆等热水用量小的场所采用分散型小型容积式电热水器供给；演员化妆淋浴用水量大的场所采用容积式电热水器若干台集中设在地下层热交换机房内，并设机械循环系统以保证干管水温，部分热水支管采用电伴热保温；餐厅厨房热水由燃气热水器直接供给。排水系统室内采用污废分流，设专用通气系统，VIP卫生间设器具通气管；室外雨污水分流。屋面雨水排水采用虹吸式雨水排水系统，6.00标高与室外连通大平台排水由建筑设置雨水暗沟收集后，利用雨水管重力流排放。

室外消防水量直接利用城市自来水及建筑西侧的人工湖取水。地下层消防水泵房内设消防水池1400立方，在屋顶设高位消防水箱，消防贮水量为36m^3。消防系统包括：室内外消火栓系统、闭式喷头的湿式系统、开式喷头的雨淋系统、防护冷却水幕系统、防火分隔水幕系统、大空间智能灭火系统、水喷雾灭火系统、七氟丙烷气体灭火系统、灭火器。

在大于12m的高大空间，按规范设置大空间智能灭火装置时，应特别关注水炮设置位置不要有遮挡，应根据保护场所的需求，选择不同的喷头，但要注意在后期配合精装设计时，观众厅天花装饰板或吸声板对射水器的遮挡。

剧场的主舞台上部有葡萄架，采用了雨淋系统；侧舞台、后舞台上部无葡萄架，但净空高度大于12M，也采用雨淋系统；主舞台和后舞台台口的防火幕采用水幕冷却防护系统；主舞台两侧台口未设钢制防火幕，均采用防火分隔水幕系统保护。

本项目根据现场情况，将餐饮废水处理装置设置在室外的后场区，非人流区的车型道下面，选用目前先进隔油技术，并结合了地形，采用池体为钢筋混凝土结构，内衬不锈钢的隔油装置，为保证北方地区冬季隔油效果，增加了电加热设备。

歌剧厅高空水炮
Intelligent Fire Extingushing System in Opera Hall

音乐厅高空水炮
Intelligent Fire Extingushing System in Musical Hall

Water Supply and Drainage Design

The professional design of water supply and drainage includes the systems of water supply, drainage, rainwater, fire and so on.

The water using of F1 and B1 are all from the municipal tap water pressure and the water using of F2 and above are from the joint water supply mode of water tank and variable speed pump. The toilet flushing water uses recycled water. Because the performance characteristics of the theater building decides the concentration and intermittent of the using water, the hot water system uses a combination of centralized and decentralized supply mode: small water consumption places such as wash basin adopt distributed small volume type electric water heater supply in the toilet; large water consumption places such as actor makeup shower use electric water heater of several centrally located in the underground heat exchanger room and set the mechanical circulatory system to ensure the stem pipe temperature and part of hot water pipes use electric heating thermal insulation; the hot water of the restaurant kitchen is supplied directly by gas water heater. The drainage system inside uses waste stream and sets special ventilation system and the VIP toilets set ventilation tubes for apparatus; the rain sewage is diversion outside. The roof rainwater and drainage use siphon rainwater drainage system. The drainage of the platform connected with outside as its elevation is 6m are collected by rain water under drain set by building professionals and use gravity flow of rainwater pipes to discharge.

Outdoor fire water directly uses the city water and the artificial lake which is on the west side of the building. The underground fire pump room equipped with a 1400m3 fire pool, and set a high-level water tank on the roof with water storage of 36m3. Fire protection system include Indoor and outdoor fire hydrant system, closed spray wet system, open sprinkler shower system, drencher for cooling protection system, fireproof water curtain system, large space intelligent fire extinguishing system, water spray fire extinguishing system, propane heptafluoride gas fire extinguishing system and fire extinguisher.

When tall spaces greater than 12m and set large space intelligent fire extinguishing system according to the standard, special attention should be paid to make sure the position of the water cannons not sheltered. Choose different kinds of nozzles according to the needs of protected location. But pay attention to the occlusion of the Imizu by audience hall ceiling decoration board or acoustic board in later hardcover design.

On the theater's main stage, there is upper part of the grape arbor, which is protected by deluge system. There is no grape beside or behind the stage, but Headroom greater than 12m, which is also protected by deluge system. The main stage after stage proscenium safety curtain protective water curtain cooling system; On both sides of the main stage proscenium Install steel fire curtain and are both protected by Fire separation water curtain system.

In this project, according to the situation, set dining wastewater treatment plant in the outdoor backcourt, below the models tract following non-traffic areas. Advanced oil separation technology is selected and Combined with the topography, cell body is a reinforced concrete structure, lined stainless steel grease trap device. In order to ensure the the northern winter grease traps effect, electric heating equipment is added.

电气设计

天津大剧院为大型甲级剧院，供电负荷等级为一级，并有部分特别重要负荷。本工程采用35kV电源供电，计算负荷9000kVA，变压器为35kV/0.4kV，2台2000kVA为一组，2台2500kVA为另一组，高低压均设母联。2台900kW柴油发电机并机发电作为第三路电源。

一级负荷采用市电双电源自切，柴油发电机作为第三路电源；剧场调光用计算机系统和计算机信息控制系统作为特别重要负荷，末端增设UPS不间断供电。舞台照明、舞台机械、贵宾室、演员化妆室和其他二级负荷采用两路市电切换供电。大剧院、音乐厅和多功能厅等人员密集场所的正常照明采用从不同变压器引的两路电源,各带一半照明负荷，交叉供电。

舞台灯光的可控硅调光控制机房和舞台机械控制机房靠近舞台负荷中心，并就地安装有源滤波器提高电能质量。舞台灯光和舞台机械控制系统与消防控制系统联动，火灾时能中断演出模式强行进入消防模式。电声设备、电视转播用电、计算机信息控制系统等对谐波敏感的设备与舞台工艺、泛光照明等设备分配在不同的变压器组，并就地配置隔离变压器以确保不受谐波的电磁干扰。预留电视转播的双切电源和室外演出灯光音响等临时用电电源。

剧院内高大空间，采用显色性佳的金卤灯配以部分LED营造星空效果。公共场所照明采用智能照明控制系统。剧院内观众厅照明控制系统与舞台灯光控制系统联动，演出时由舞台灯控集控。泛光照明采用内光外透，LED光源节能。应急疏散照明采取智能应急疏散照明系统，消防时根据实际情况指导人员按最佳路线逃生。

火灾自动报警采用控制中心火灾报警系统（FAS）。剧院内16~26m舞台高大空间选用空气采样报警系统，前厅高达18m的挑空采用红外线探测器，结合水炮探测灭火系统。以上系 统均与消防自动报警及联动控制系统形成集控。

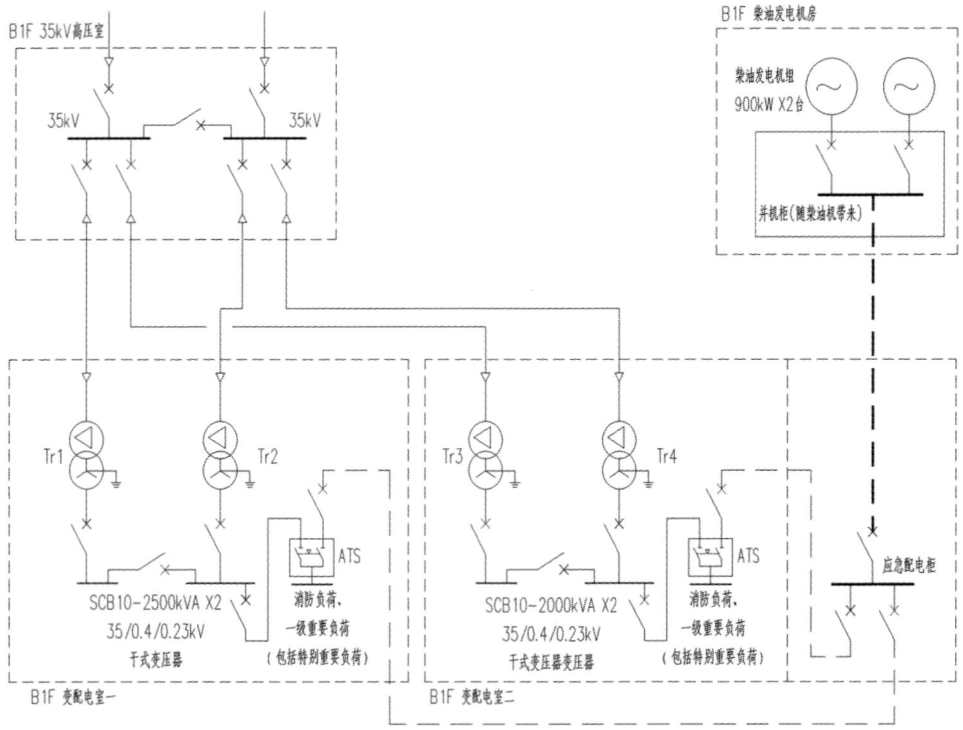

高低压系统图
High Voltage System

Electrical Design

Tianjin Grand Theatre is a large class-A theater. Load power supply classification is first grade and part is vital grade. It takes 35kV power supply for the project and calculated load is 9000kVA. Four 35kV/0.4kV transformers were set, two 2000kVA as a group, two 2500kVA as another group, with connection on both high and low voltage. In addition, two 900kW diesel generators were set as the third power.

It takes automatic switching double power sources for first grade load ,with diesel generator as the third power. As vital load in first grade load, the computer and information control system for theatre dimming was added UPS at the load end. The load of stage lighting, stage machinery, VIP room, actor dressing rooms and other second grade load was used two switching power supply.

The normal lighting load of theatre, music hall and multifunction room and other personnel-intensive places was supplied by two powers from different transformers. And each power loads half of the lighting, crossed.

The triac dimmer control room of the stage lighting and stage machinery control room were set near the stage load center and active power filter was installed on-site to improve power quality.

The stage lighting and stage machinery control system was integrated with fire control system to interrupt presentation mode and force into the fire mode once fire. Sensitive equipments for harmonic such as audio equipment, television relay, computer information control system and so on were distributed in different transformer groups with stage craft, floodlight lighting equipments. And isolation transformers were deployed on-site to ensure no harmonic electromagnetic interference. Duplicate supply for television relay and temporary power supply for outdoor performance audio and lighting and so on was reserved.

Good color rendering gold halogen lamps with part LED lights were used to create a starry sky effect in the tall space within the theater. Intelligent lighting control system was used in public area. The auditorium lighting control system was integrated with stage lighting control system and controlled centralized by the stage lighting control system When the performances. Floodlight uses energy-saving LED lights, outside through the inner light. It takes intelligent emergency evacuation lighting system for emergency evacuation lighting and it can provide the best route guidance based on the actual situation when the fire. The form of automatic fire alarm system is control center fire alarm system (FAS). Air sampling alarm system was used in 16 ~ 26m tall stage space within the theater and infrared detectors combined with water cannons detect fire extinguishing system was used in up to 18m air pick of the antechamber. The above systems and automatic fire alarm and control system were connected to form a centralized control.

大剧院夜景泛光照明
Night View of Grand Theater

博物馆 Museum

华南理工大学建筑设计研究院
Architectural Design Research Institute of SCUT

建筑设计

在天津博物馆的设计中我们以贯穿博物馆，联系各个展厅的宽敞、宏大的公共大厅“世纪之窗”为空间主题，再现天津的悠久历史和重要地位。

建筑主体外表面采用625x1250mm的黄锈石外墙挂板，厚重有力的条形肌理形成博物馆的鲜明个性。裙房基座采用相同的石材肌理，与主体统一，而需要自然通风采光处，则以石材百叶形式处理。贯穿建筑的公共空间“世纪之窗”以雨点花锈蚀紫铜板作为外饰面，强化这一核心空间的形象，并形成深远的历史感。

博物馆北向主入口以6层逐级放大的叠涩象征天津设卫建城600年悠久历史，引导公众由开阔的滨水广场进入博物馆时光隧道大厅。由此向南纵向逐级上升，层层叠叠，形成强烈的纵深感，仿若时光隧道。大厅入口宽度30米，总长85米，平均高度14米，利用层层提高的台阶，结合自动扶梯，依次连接古代、近代、现代展厅，带领公众游历天津的文明和历史发展。时光隧道大厅是天津博物馆最主要的空间，在不同标高的展示平台设置6组主体群雕，象征天津600年的发展历程。时光隧道大厅墙地面以浅黄石材为主，形成整体简洁的展示背景空间，结合门套、铜浮雕等铜饰构件，体现大气、厚重的博物馆气质。

“世纪之窗”在南端横向展开成宏大的未来大厅，充分展现天津文化中心和城市景色，预示着天津的美好未来。未来大厅总宽105米，进深18米，净高14.5米，是以展示、休息、集散为主的综合性公共空间。该空间设有宽105米、高14.5m的巨型通透景观玻璃幕墙面向市区。在此，城市景观一览无遗，回望时光隧道，则整个大厅尽收眼底。将外墙的紫铜材质延伸至未来之窗大厅室内墙面、顶面，实现了室内外材料的延伸感，结合良好的采光，形成了恢弘的金色大厅。

博物馆展陈设计通过时光隧道大厅，将古代、近代和当代等历史类陈列厅连为一体，依时间顺序层层上升，喻示天津悠久的历史，而终结处的全景式未来大厅则象征广阔的未来。绘画、书法、玉器等文物类陈列厅在博物馆顶层水平展开，表现天津丰硕的文明与文化。

时光隧道
Time Tunnel

摄影：姚力
Photographer：Yao Li

未来之窗
Window of the Future

Architectural Design

In the design of Tianjin Museum, we represent the long history and important position of Tianjin by taking the spacious and grand public hall, i.e. the "Window of the Century" throughout the museum for contacting the various exposition halls as the space theme.

The external surface for the main body of the building is constructed by hanging palo gold plates on the external wall, wherein the thick and strong strip-shaped texture forms the distinct character of the museum. The podium is constructed with the same stone texture and uniformed with the main body, and the position with the requirements of natural ventilation and lighting is treated as a stone shutter. For the public space, i.e. the "Window of the Century" throughout the building, the external finish is a copper plate corroded as rain flower, so that the image of the core space is strengthened and a profound sense of history is formed.

The north main entrance of the museum is featured with a gradually enlarged 6-layer corbel, which symbolizes the long history of 600 years since the building of Tianjin and guides the public to enter into the museum from the wide waterfront plaza.

The public hall, longitudinally and gradually risen and piled to form a strong sense of depth, is seemed like a time tunnel The width for the entrance of the hall is 30m, the total length is 85m, the average height is 14m, and the hall is orderly connected to the ancient, recent and modern exhibition halls to show the public around the civilization and historical development of Tianjin through the steps, which are raised layer upon layer and combined with an escalator. The hall of time tunnel is the main space of Tianjin Museum, in which the exhibition platforms of different elevations are equipped with 6 sets of subject group sculptures, which symbolize the development history of Tianjin in the modern history of 600 years. The wall and ground for the hall of time tunnel, mainly constructed with light-yellow stone and combined with door pocket, copper relief and other copper ornaments under the premise of meeting the wholly neat exhibition background space, reflects the meteoric and heavy temperament of a museum.

In the southern end, the "Window of the Century" is transversely expanded into a grand future hall to fully show the cultural centre and urban landscapes of Tianjin and indicate the excellent future of Tianjin. The total width of the future hall is 105m, the depth is 18m, the headroom is 14.5m, and the future hall is a comprehensive public space based on exhibition, rest and distribution. The space is featured with a giant and transparent glass landscape curtain wall (105m*14.5m) facing to the urban area, and therefore the urban landscape is in full view, and the exhibition image of the whole hall is also in full view while looking back the time tunnel. Due to the good lighting and spreading feel of indoor and outdoor materials, a generous golden hall is formed by extending the red copper of the external wall to the wall and roof within the hall of the Winder of the Future.

For the exhibition design of the museum, the ancient, recent, modern and other history exhibition halls, connected through the hall of time tunnel and raised layer upon layer, indicate the long history of Tianjin, and the panoramic hall in the end point symbolizes the great future. The painting, calligraphy, jade and other cultural relic exhibition halls, horizontally arranged on the top floor of the museum, reflect the rich civilization and culture of Tianjin.

夜景 Night View

东北角透视 Perspective of Northeast Coast

北透视
North View

东南透视
Southeast View

西北透视
Northwest View

夜景
Night view

东北角透视
Northeast View

东立面
East View

未来之窗
Window of the Future

天窗、平台、铜门
Skylight, Platform and Copper Gate

庭院
Courtyard

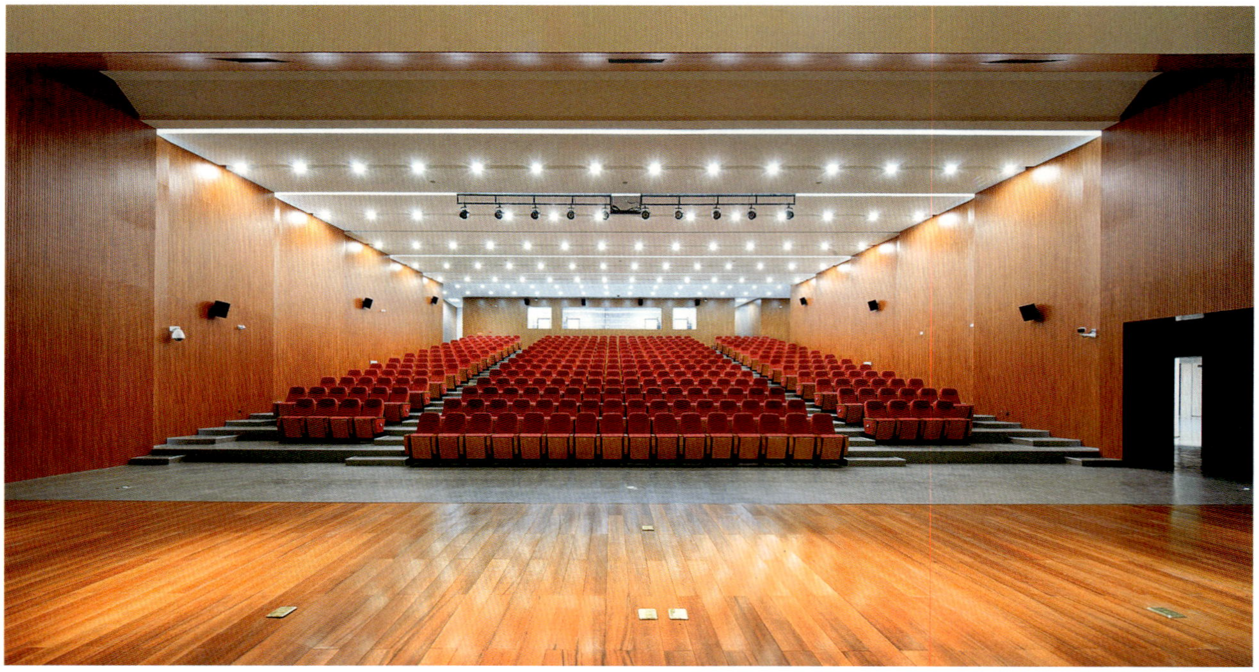

报告厅
Lecture Hall

时光
Time Tu

平面图 Plan

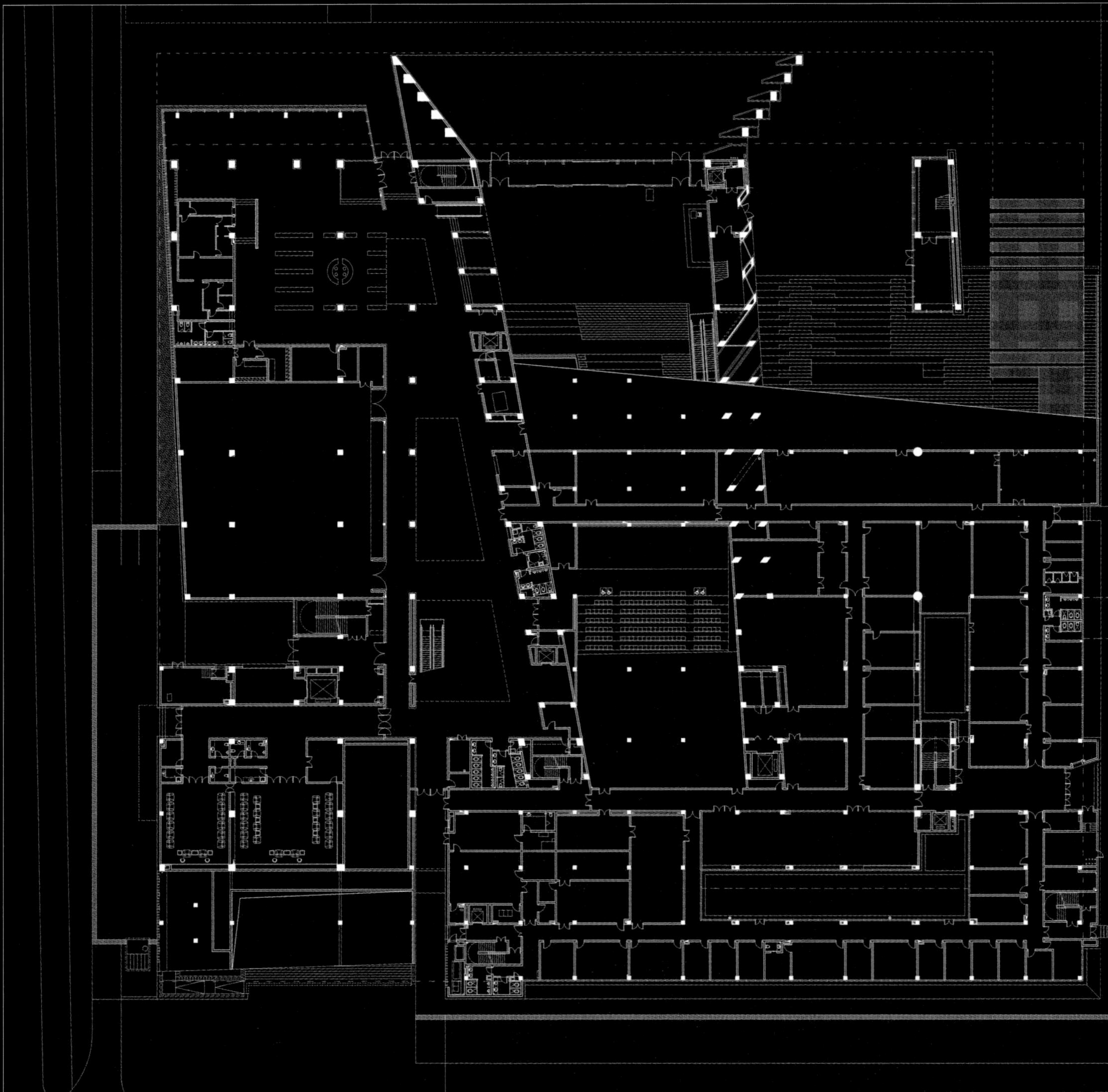

首层平面图 Plan of First Floor

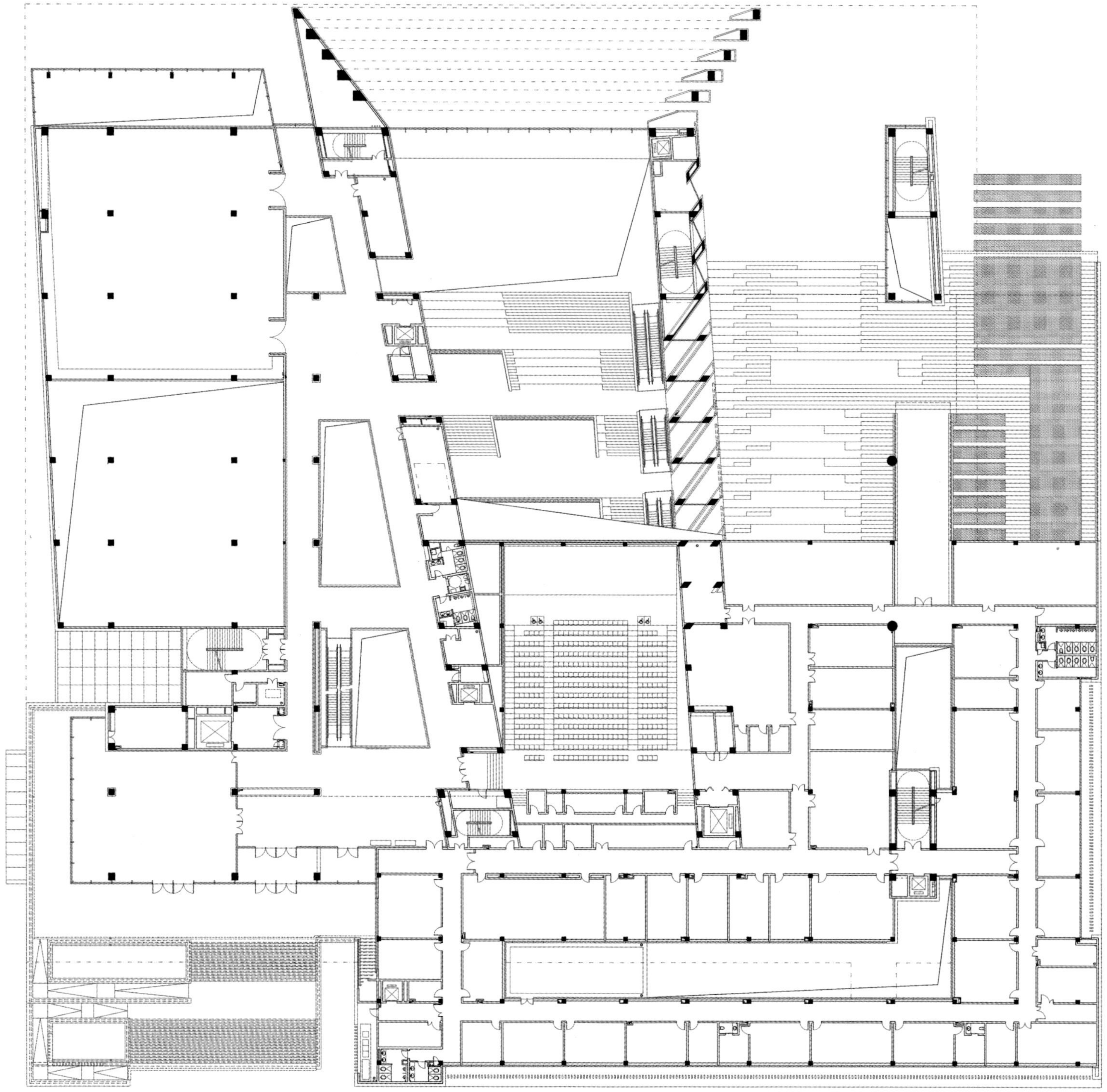

二层平面图 Plan of Second Floor

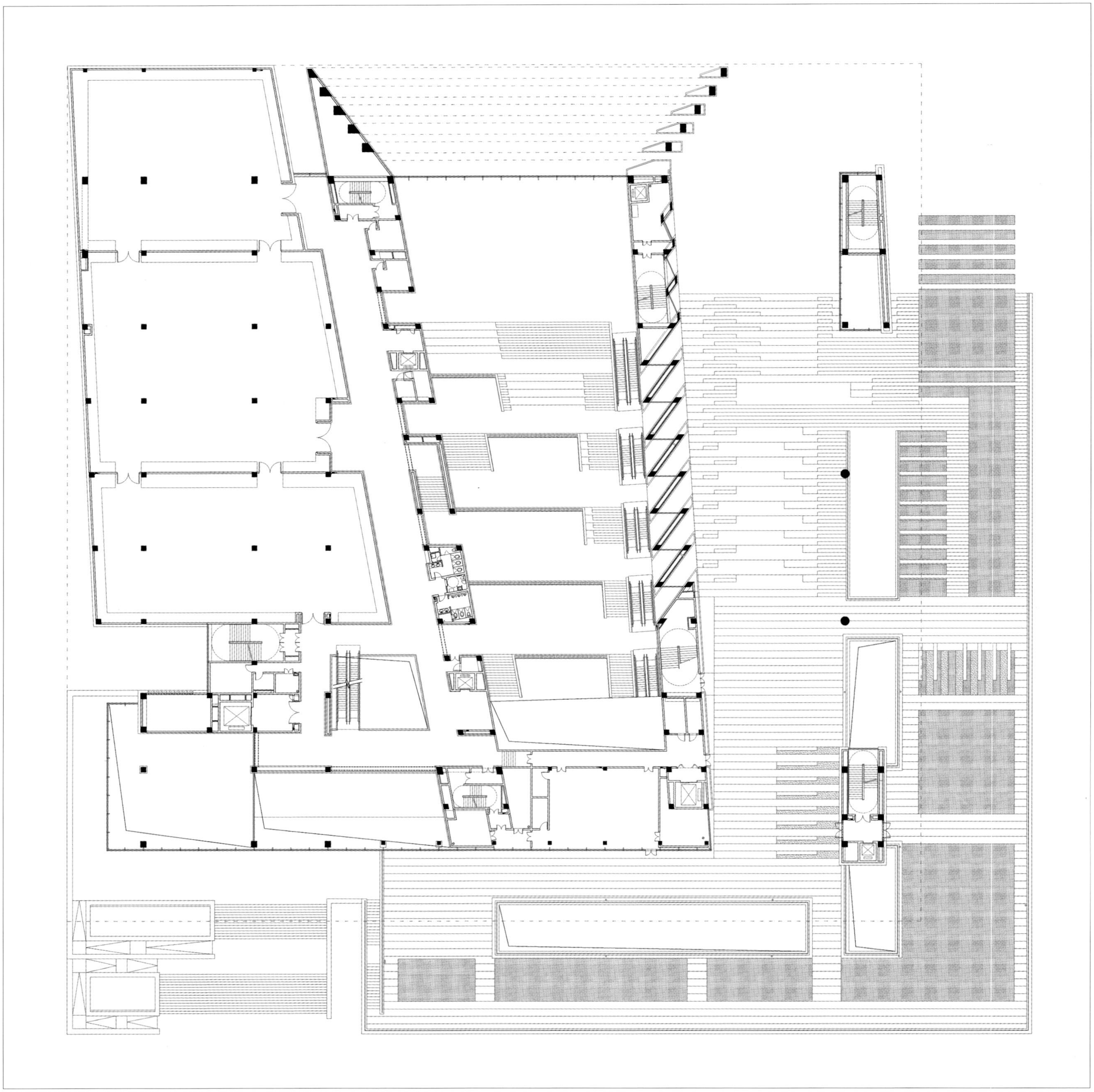

三层平面图 Plan of Third Floor

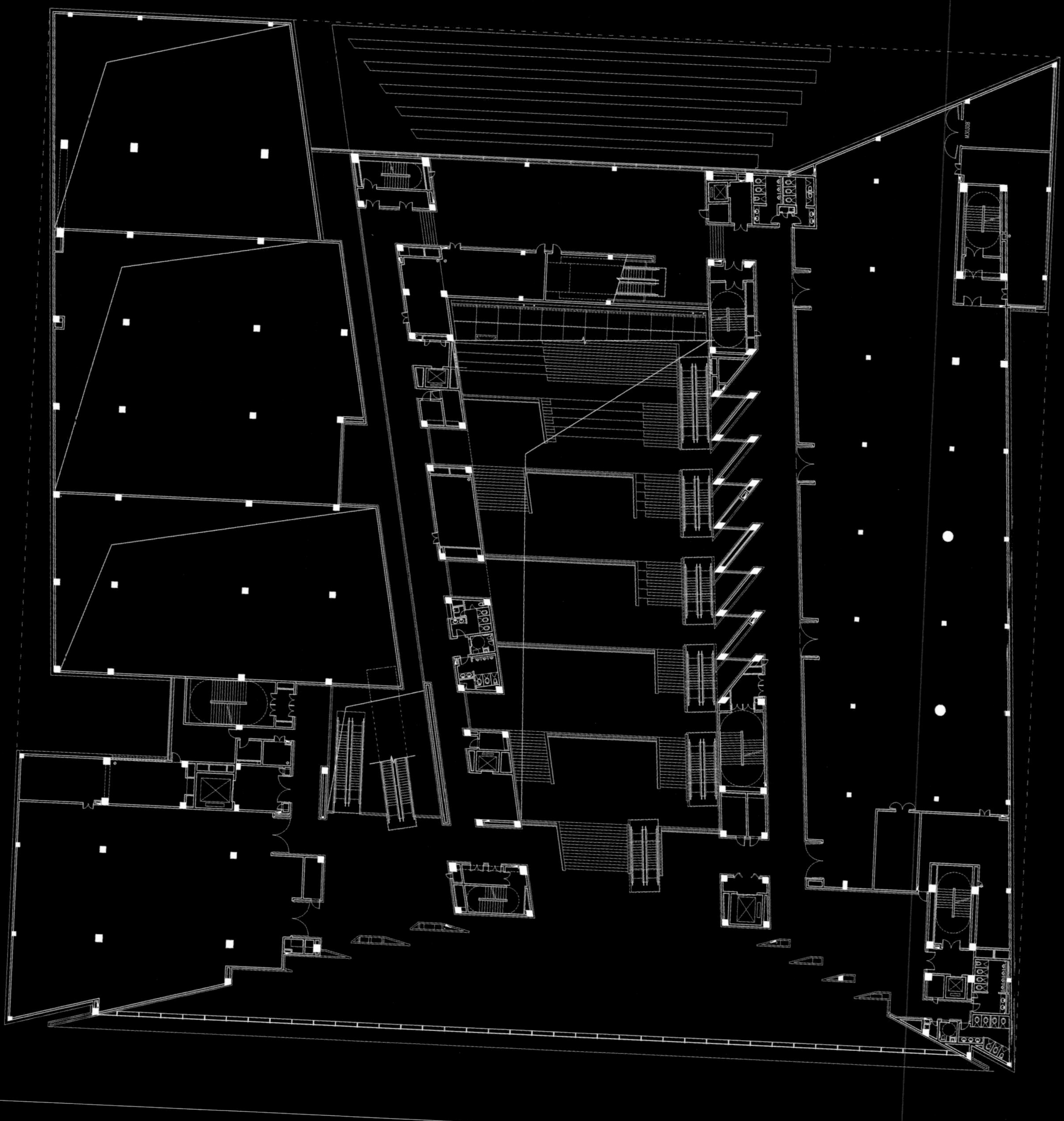

四层平面图　Plan of Fourth Floor

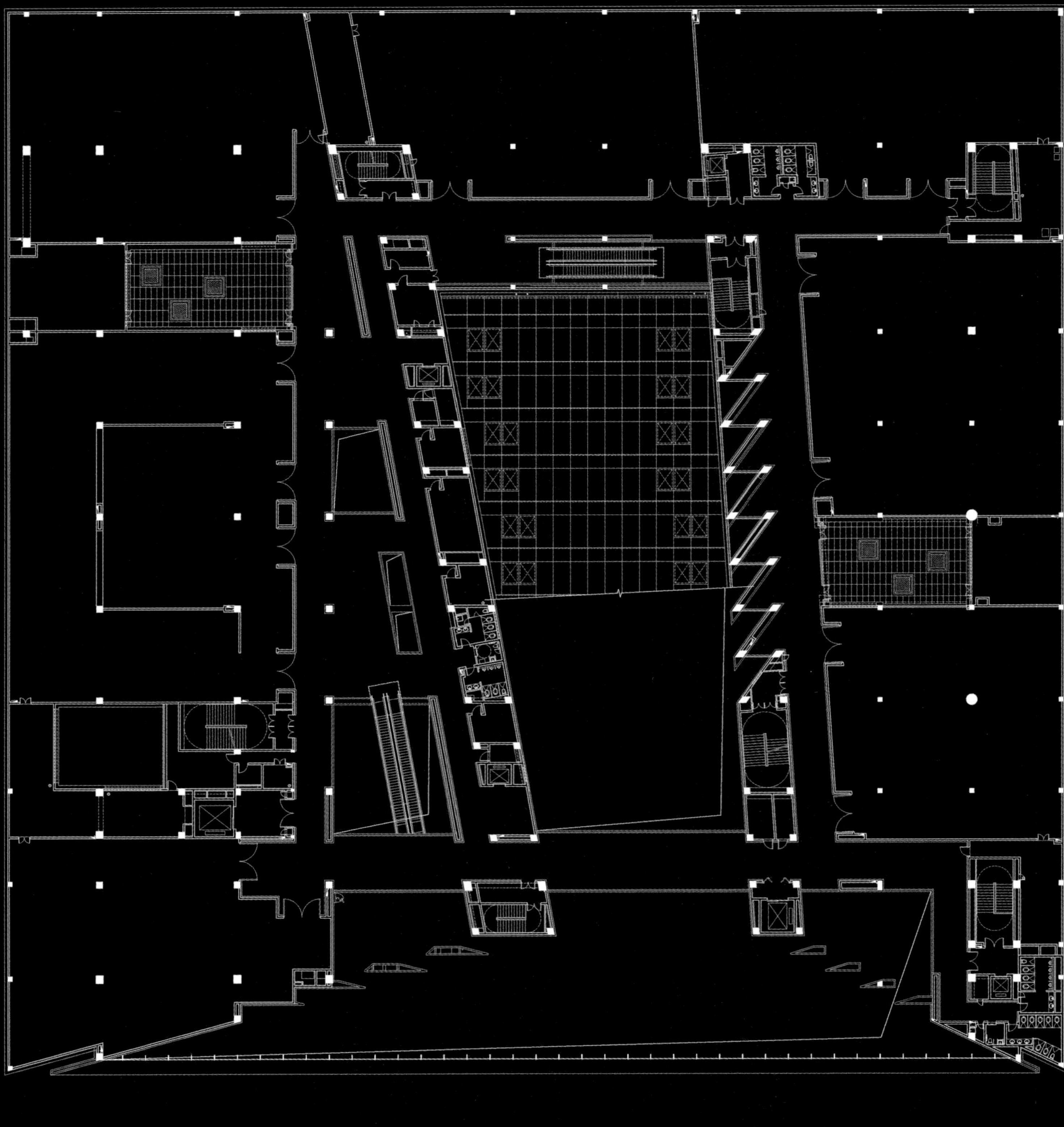

平面图　Plan of Fifth Floor

剖面图 Section

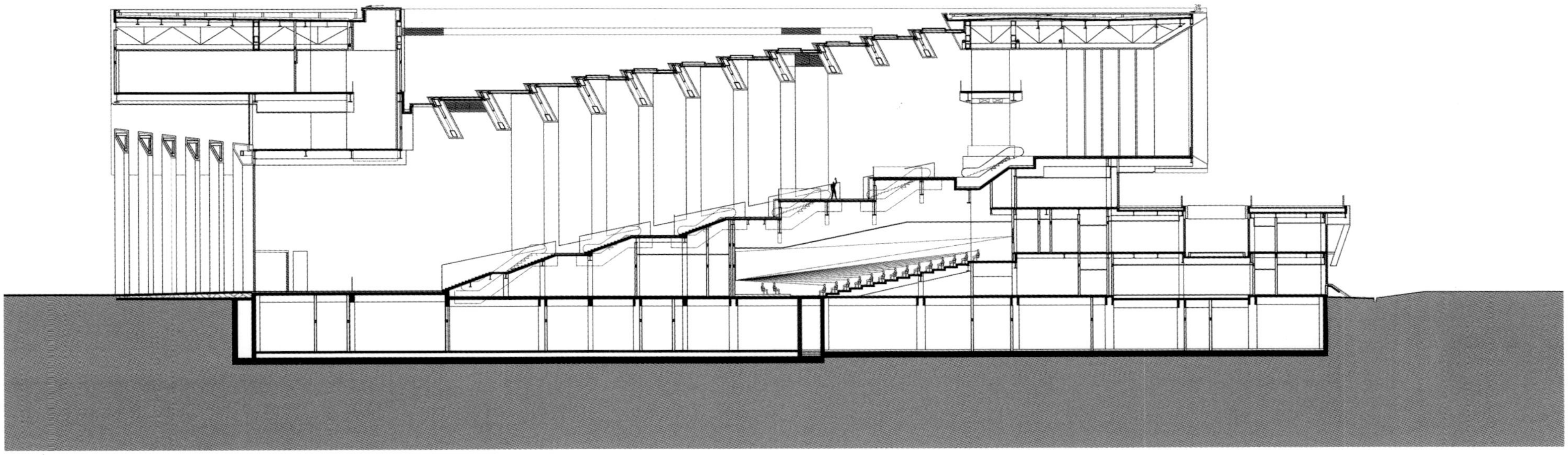

1 – 1 剖面图
1-1 Section

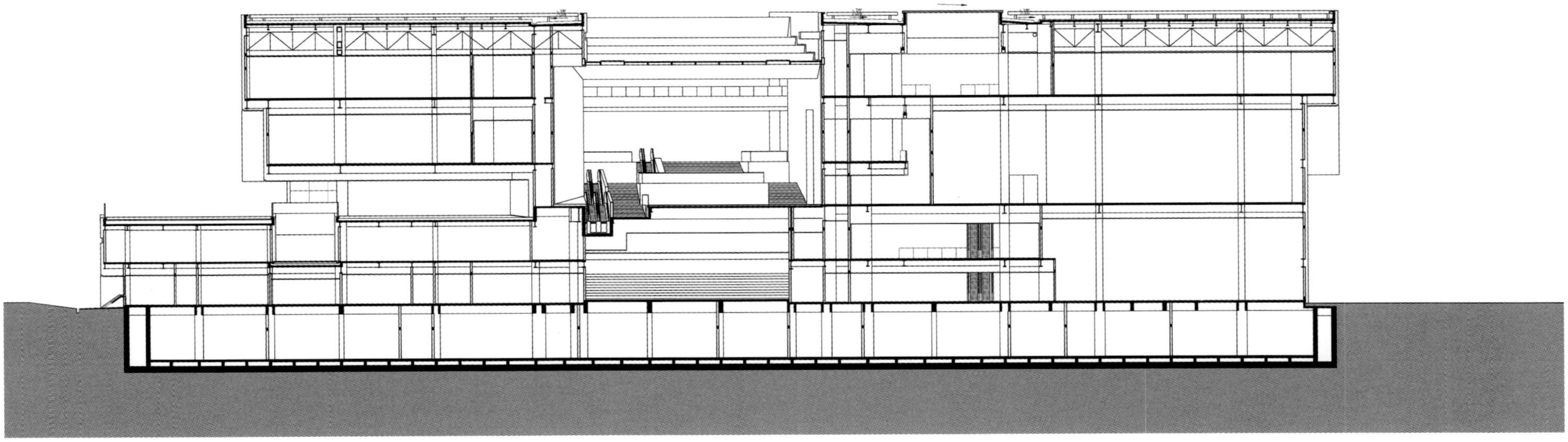

2 – 2 剖面图
2-2 Section

建筑构造 幕墙剖面大样
Architectural Construction Sectional Detail of Curtain Wall

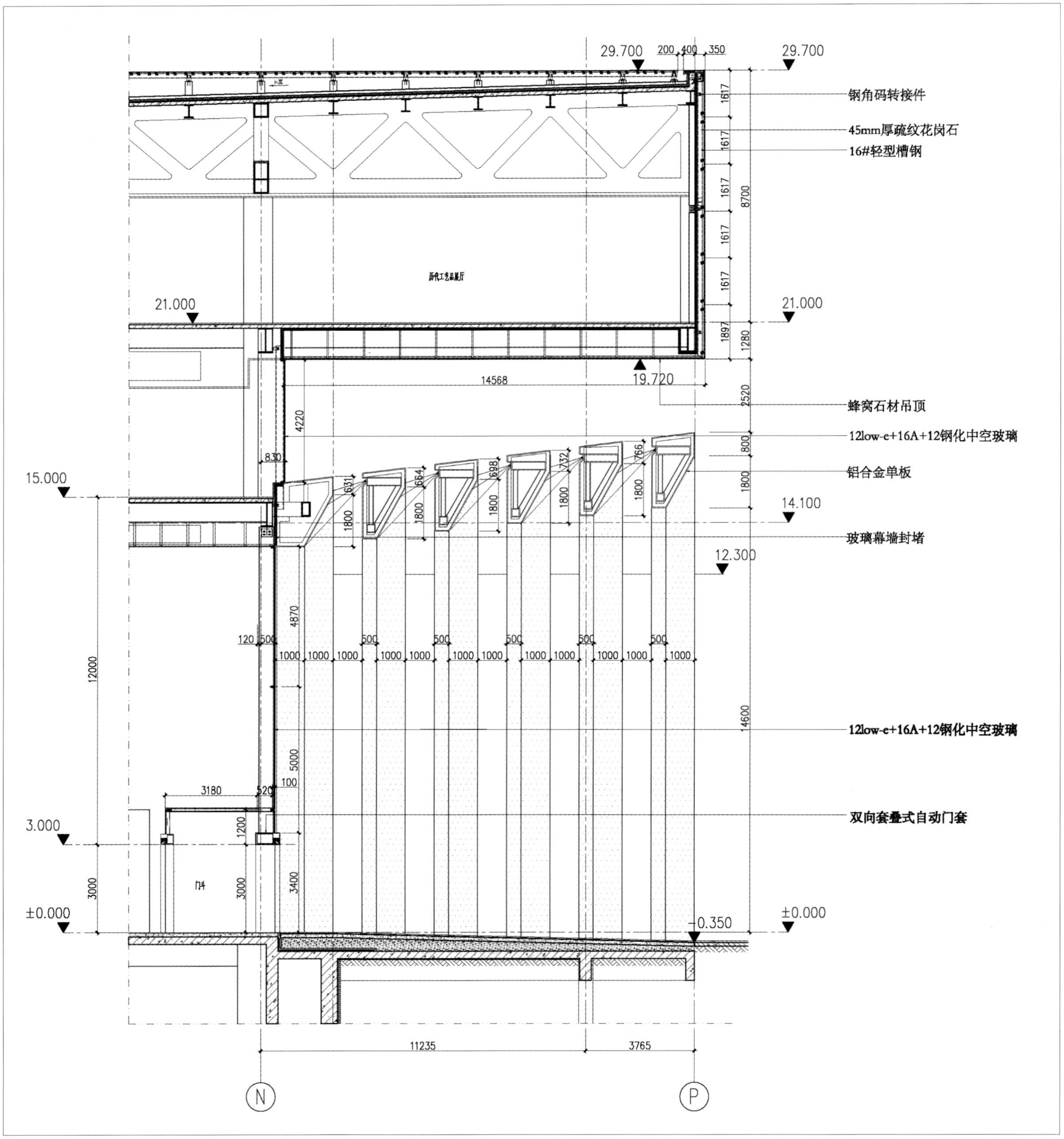

主入口幕墙平面详图
Detailed Plan for Curtain Wall of Main Entrance

主入口六重门
Six-layer Corbel of Main Entrance

建筑构造 石材幕墙大样
Architectural Construction Detail of Stone Curtain Wall

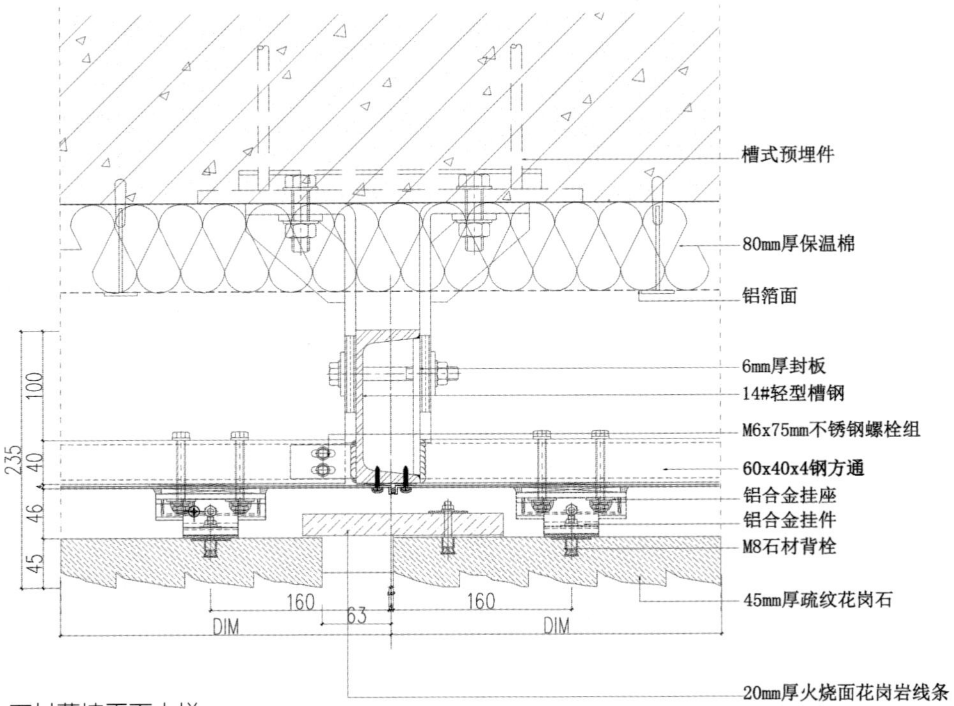

石材幕墙平面大样
Detailed Plan of Stone Curtain Wall

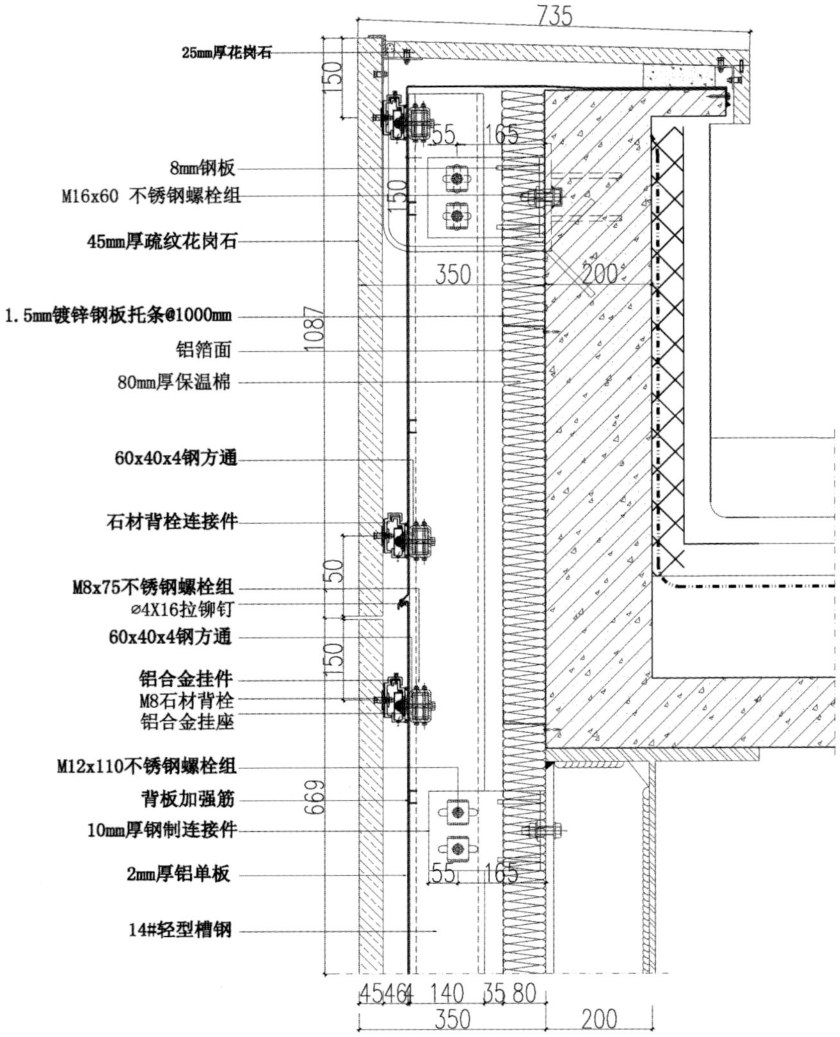

石材幕墙剖面大样
Sectional Detail of Stone Curtain Wall

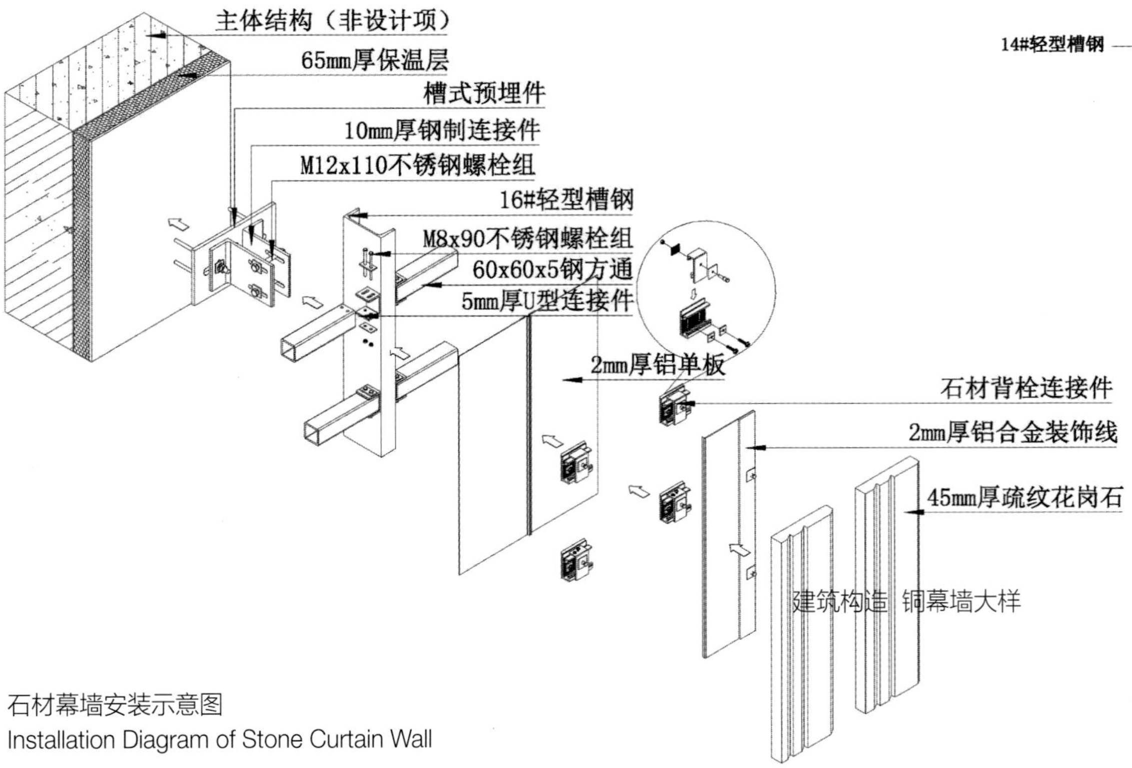

石材幕墙安装示意图
Installation Diagram of Stone Curtain Wall

石材肌理
Stone Texture

建筑构造 铜幕墙大样

Architectural Construction Detail of Copper Curtain Wall

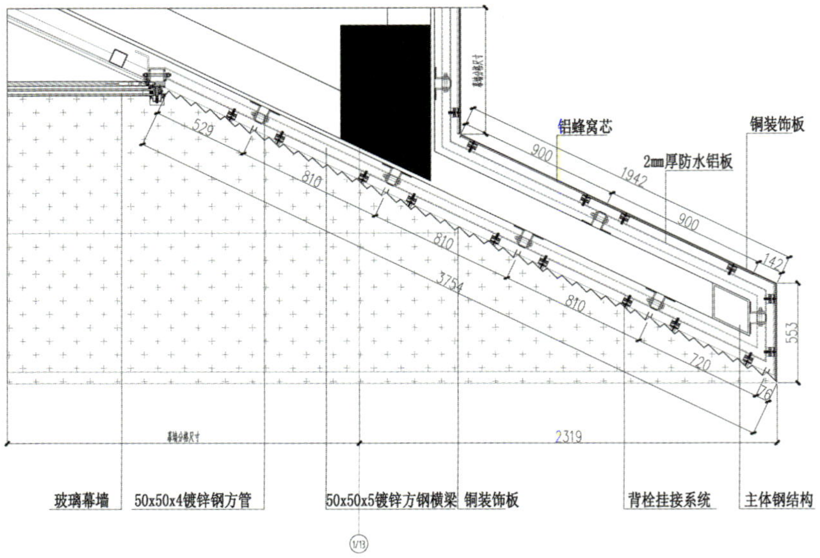

铜幕墙平面大样
Detailed Plan of Copper Curtain Wall

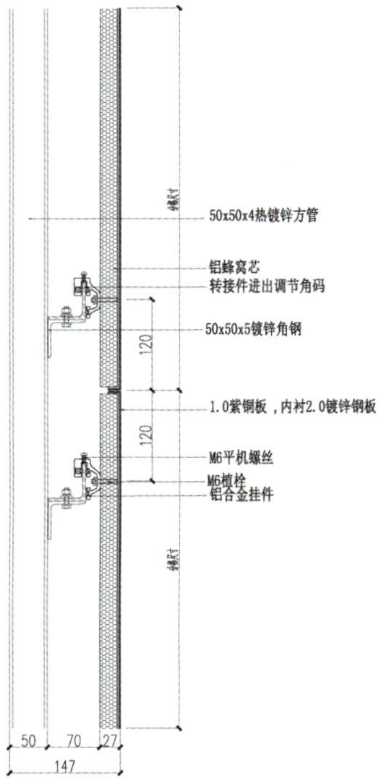

折板铜幕墙大样
Detail for Copper Curtain Wall of Folded Plate

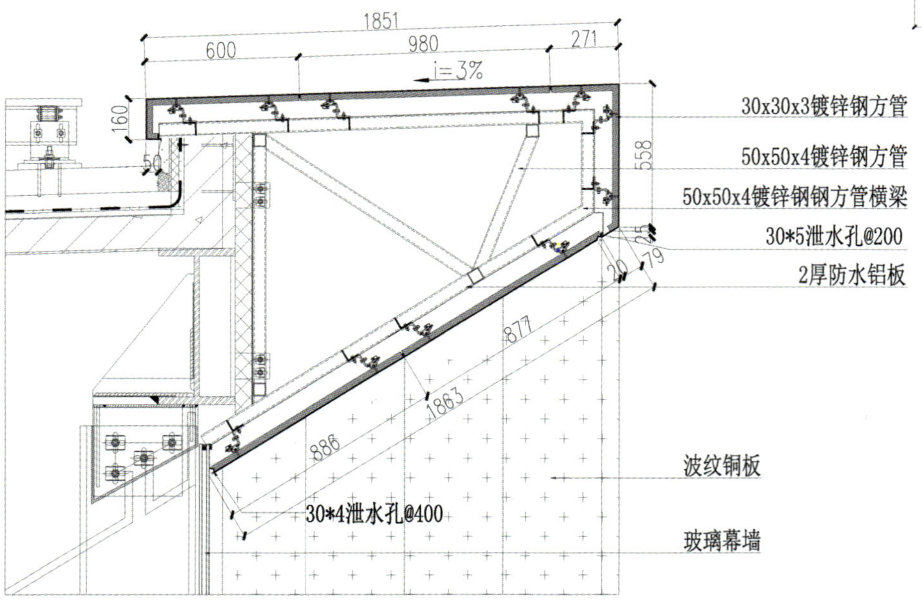

铜幕墙剖面大样
Sectional Detail of Copper Curtain Wall

未来之窗铜幕墙 Copper Curtain Wall of Window of the Future

建筑构造 内装
Architectural Construction Interior Decoration

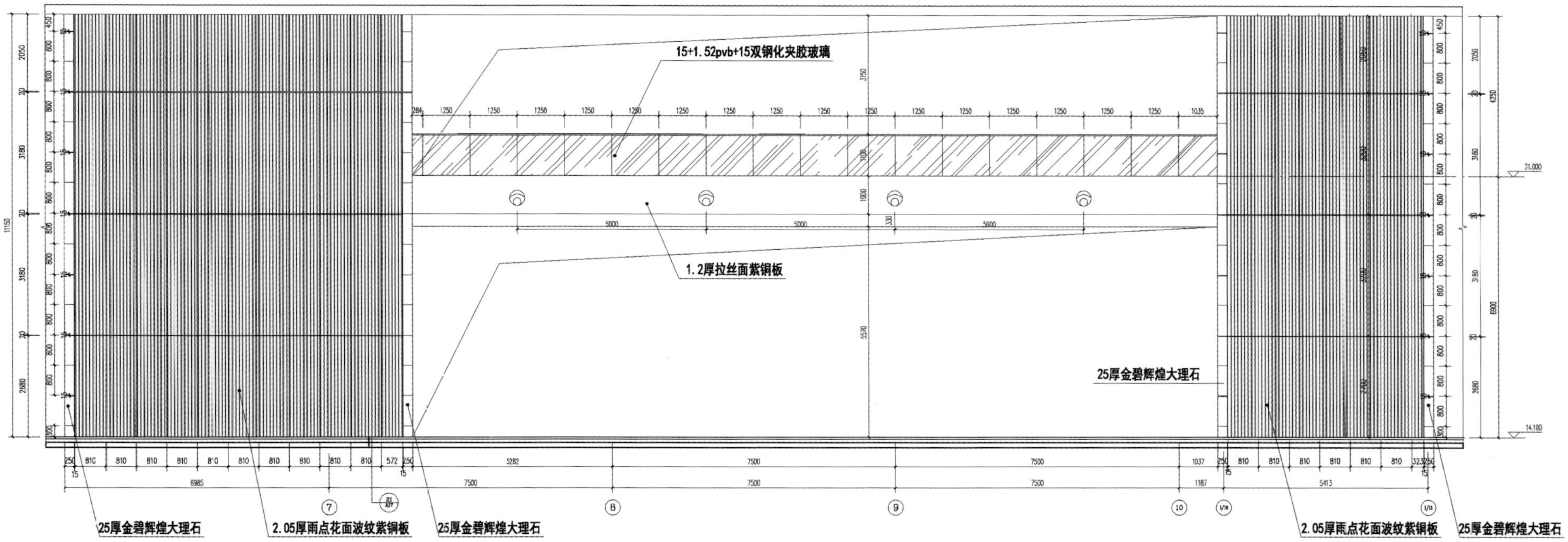

未来之窗廊桥立面图
Elevation for Covered Bridge of Window of the Future

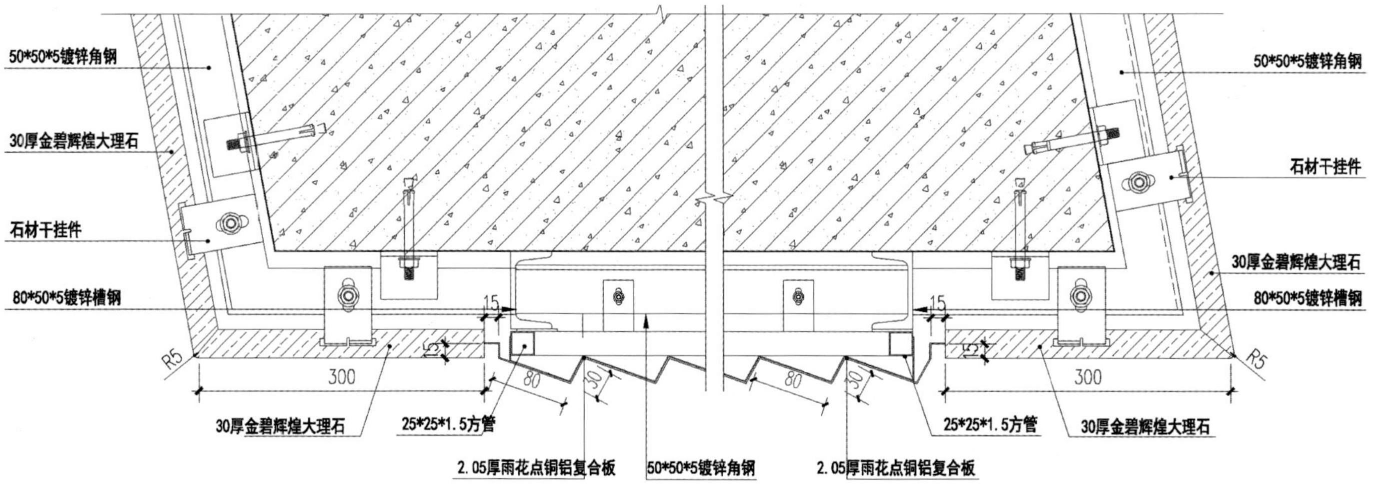

波纹铜板与石材收口节点
Convergent Node of Corrugated Copper Plate and Stone

建造 施工现场
Construction Site

15 米悬挑结构
Cantilever Structure of 15m

30 米跨度时光隧道
Time Tunnel with the Span of 30m

结构设计

本工程采用钢框架-支撑结构体系。利用电梯井、楼梯间位置对称设置钢支撑框架，与框架柱、顶层桁架一起作为主要的抗侧力结构。采用矩形或圆形钢管混凝土柱作为主要竖向主要承重构件。绝大部分钢管混凝土柱直接落至地下一层底板面，并在地下室段外包200mm厚钢筋混凝土。对应地上支撑框架位置，地下室部分布置400mm厚的钢筋混凝土墙体。二层以上楼盖一般采用密肋钢梁-混凝土板组合楼盖体系。地下室顶板采用普通钢筋混凝土梁板结构。矩形钢管柱在地下室部分采用外包混凝土做法，与钢管柱相连的主梁采用双梁，从钢管柱两侧通过。

本建筑平面凹凸尺寸接近相应边长的30%，楼板凹凸不规则，楼板有效宽度小于50%，开洞面积大于30%，错层大于梁高，楼板局部不连续，削弱较大。为了满足整体结构在大震中的表现，结构楼板应有足够的刚度和承载力，其重要性甚比其他主要构件。因此楼板按中震弹性进行设计。为保证结构整体性，各层楼盖采用双层双向钢筋网抵抗楼板应力，支撑框架附近楼板应力较大处加强钢筋配置。

建筑屋盖四周设有大悬挑桁架结构，最大出挑15m，并下挂一层或至两层楼板，这不仅造成结构竖向质量分布不均匀，还使楼盖的转动惯量增大。不但造成竖向质量分布不均匀，还使得楼盖的转动惯量增大。因此，沿出挑根部四周尤其是建筑物的四个角落布置钢支撑最能抵抗扭转作用。为了保证扭转的规则性，钢支撑尽量进行对称布置。

Structural Design

The project uses a steel frame - support structural system. The steel support frames are symmetrically arranged according to the position of the elevator shaft and the staircase, and jointly used as main lateral resisting structures with the frame columns and top trusses. The main vertical load bearing components are rectangular or circular concrete-filled steel tube columns, most of which are directly fallen into baseboard of the first underground floor and covered with the reinforced concrete of 200mm thick in the basement section. Corresponding to the position of ground support frame, the basement is partially equipped with the reinforced concrete wall of 400mm. Generally, the floor system above the second floor is a combined floor system of multi-ribbed steel beam and concrete plate, the top plate of the basement is an ordinary reinforced concrete beam slab structure, the part of the rectangular steel tube column in the basement is externally covered with concrete, and the main beam connected to the steel tube column is a double beam passing through the two sides of the steel tube column.

Due to the requirement of large space for the overtravel of the museum, the bump size of the plane is close to 30% of length of the corresponding side, the bump of the floor slab is irregular. In addition, the effective width of the floor slab is less than 50%, the opening area is more than 30%, the split level is more than the height of the beam, the floor slab is partial discontinuous, and the weakening is relatively obvious as the courtyard, i.e. the "Time Tunnel" is ascended step by step. In order to realize the performance of the overall structure in the large earthquake, the floor slab of the structure shall have sufficient stiffness and bearing capacity, and the importance thereof is comparable to other major components, and therefore the floor slab shall be designed as intermediate earthquake. In order to ensure the integrity of the structure, the floor system of each floor shall be provided with a double-layer and two-way steel mesh to resist the stress of the floor slab, and the reinforcement configuration for the place of the support frame close to the floor slab and featured with a larger stress shall be strengthened.

Based on the form, the periphery of the roof is provided with large cantilevered truss structures, wherein the largest projection is 15m, the lower part of the structure is provided with one or two layers of floor slab by hanging, and therefore the vertical mass distribution of the structure is uneven, and the rotation inertia of the floor system is increased. Therefore, the periphery of the projected root part, especially the four corners of the building are provided with steel supports with most resistance to the torsion in the structure arrangement process. In order to ensure the regularity of torsion, the steel supports shall be symmetrically arranged as much as possible.

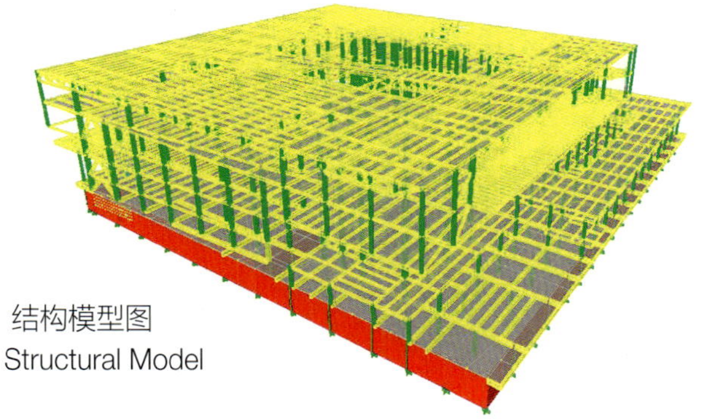

结构模型图
Structural Model

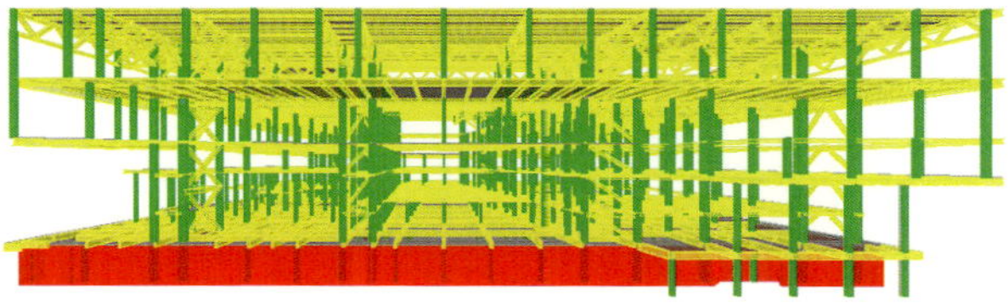

结构模型图
Structural Model

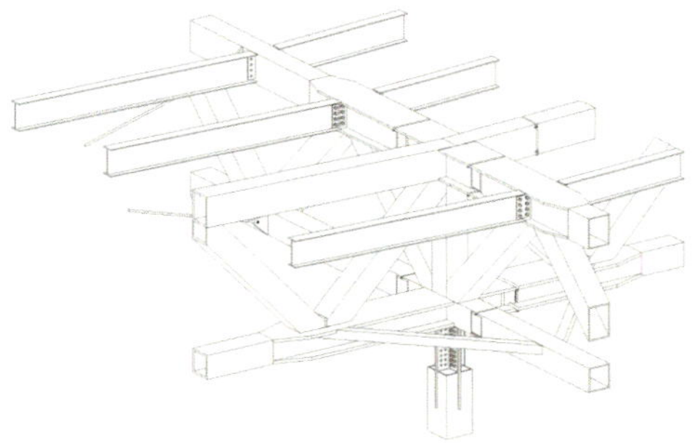

屋面桁架与吊柱连接大样
Detail for the Connection between Roof Truss and Davit

暖通空调设计

本工程采用区域集中供冷和供热系统，在地下室设置板式换热器间，选择2台换热量为3840KW的板式换热机组，冬季将区域热力管网47℃/38℃一次热水换热得到43℃/36℃二次热水供空调供热用，夏季将区域冷水管网3℃/12℃冷水二次换热得到6℃/13℃的空调冷水。空调冷水系统采用一次泵变流量系统，根据不同分区的使用要求划分多个水系统环路，分别满足各区域冬季供热、夏季供冷需求。

考虑到博物馆恒温恒湿空调系统的不间断要求，本工程同时设立6套模块式风冷冷水机组LJ-1~6，单台制冷量65KW；设立6套模块式空气源热泵机组LJ-7~12，单台制冷量62KW、制热量67KW，以在区域冷、热水管网检修时提供备用的冷热源供恒温恒湿库房用。备用冷热源系统采用一次泵定流量系统。为满足不同功能房间的室内热舒适要求，大空间（如大厅、报告厅、展厅等）舒适性空调采用一次回风全空气低速空调系统。北侧入口门厅采用旋流风口下送和可调角度的电控球形喷口下送相结合的送风方式。小房间（如办公室等）采用风机盘管加新风系统，气流组织采用侧送上回或上送上回的方式。

共享大厅室内高达25m，并且入口朝北，如何解决其冬季采暖问题是设计面临的难点。为了解决空调采暖导致室内温度梯度大、能耗高、空调送风难度大的不足。本设计引入节能环保的低温辐射采暖形式。为了解决展厅高大空间冬季送风难度大，内区无热负荷的难题，空调风系统按照内外区分别设置送风支管，设置电动阀，冬季可以关闭内区风管，提高外区送风风速，改善热风上浮的现象。

HVAC Design

The project uses regional centralized cooling and heating systems, specifically the basement is equipped with a plate heat exchanger room and two units of plate heat exchanger with the heat exchange capacity of 3840KW. In winter, the primary hot water of 47℃/38℃ of the regional heating pipe network is subjected to heat exchange to obtain the secondary hot water of 43℃/36℃ for the heat supply of air conditioner. In summer, the cold water of 3℃/12℃ of the regional cold water pipe network is subjected to heat exchange to obtain the air conditioner cold water of 6℃/13℃. The cold water system of the air conditioner is a variable primary flow system, in which a variety of water system loops are divided to meet the requirements of heating in winter and cooling in summer of the various regions according to the different regions.

Due to the uninterrupted requirements for the constant temperature and humidity air conditioning system of the museum, the project is simultaneously provided with six sets of modular air-cooled chillers, i.e. LJ-1 to LJ-6, wherein the cooling capacity of a single unit is 65KW; the project is provided with six sets of modular air source heat pumps, i.e. LJ-7 to LJ-12, wherein the cooling capacity of a single unit is 62KW and the heating capacity of a single unit is 67KW, and the units are used for providing standby cold and heat sources for the storehouse with constant temperature and humidity while inspecting the cold and hot water pipe networks. All the standby cold and heat sources are fixed primary flow system. In order to meet the requirements of indoor thermal comfort of different functional rooms, the comfortable air conditioner of the large space (such as hall, lecture hall, exhibition hall and etc.) is a primary air returning full-air low-speed air conditioning system. The entrance hallway of the north side is featured with a combined air supply mode of swirl diffuser and an electric spherical nozzle with adjustable angle. The small room (such as the office and etc.) is provided with a fan coil and a fresh air system, in which the airflow organization is carried out by means of side supply and upper return or upper supply and upper return.

For the sharing hall, the indoor height of is 25m, the entrance is faced to the north, and how to realize the heating purpose in winter is the difficulty of design. In order to solve the problems of large indoor temperature gradient, high energy consumption and large difficulty of air supply caused by the air conditioning heating, an energy-saving and environmentally friendly low-temperature radiant heating mode is introduced in the design.

In order to solve the problem that the air supply for the hall with a high and large space is difficult in winter and the internal area does not have a heat load, the conditioner air duct system is equipped with air supply branches according to the internal and external divisions, and equipped with a electrically operated valve to close the internal air supply pipe to improve the air supply speed of the external area and improve the phenomenon of hot air floating in winter.

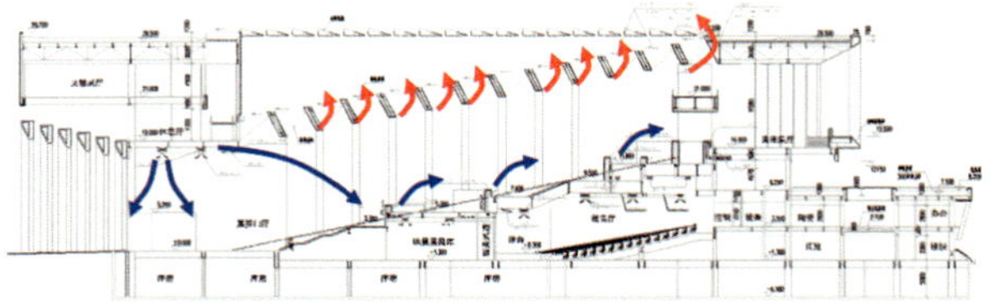

入口门厅及共享大厅气流组织方式
Airflow Organization Mode of Entrance Hall and Atrium

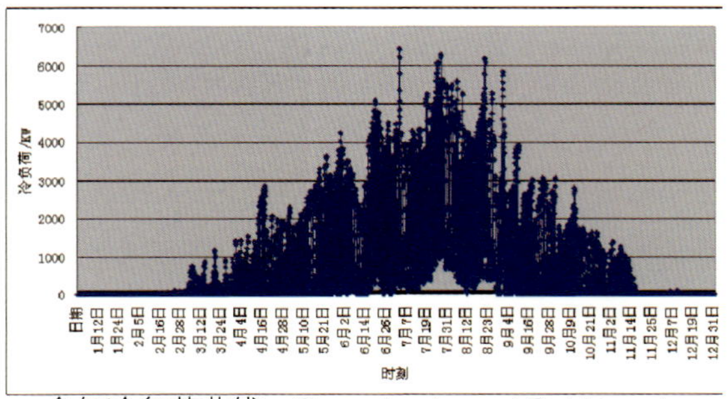

全年冷负荷曲线　Curve for Annual Cold Loads

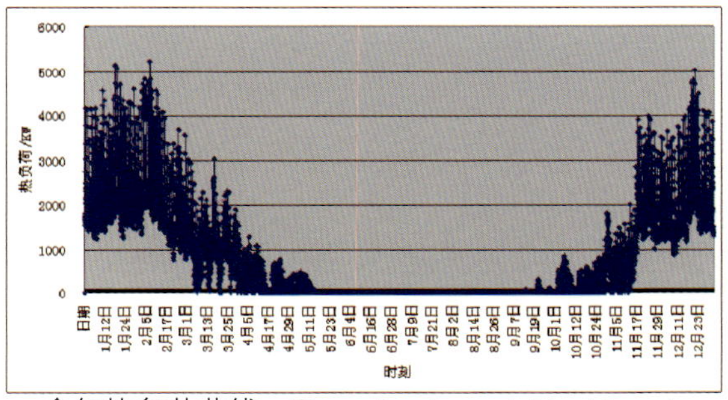

全年热负荷曲线　Curve for Annual Heat Loads

给排水设计

本建筑采用两种水源供水，室外绿化、室内冲厕用水采用中水，由市政中水管提供水源；其他用水点均采用城市自来水供水。本项目生活给水2层及以下用水点采用市政压力直接供水；2层以上用水点采用二次加压供水。中水系统采用无负压给水设备供水。室内采用污废合流方式，设置专用通气管；室外排水采用雨污分流方式。生活污水在室外经2座75m3化粪池处理，餐厅厨房污水经隔油池处理后排入市政污水管。屋面雨水设计重现期取10年，并设置安全溢流口。屋面采用压力流（虹吸）排水系统；平台雨水采用重力流排水系统，由雨水管道收集排至室外检查井或雨水沟。

室外消防用水由市政消火栓供水。室内消火栓系统为临时高压给水系统。在－1层消防水泵房设置消火栓系统气压给水设备，消防加压采用主泵、稳压泵、气压罐型式。由出水管上压力传感器自动控制稳压泵和主泵的启停。全建筑均设置室内消火栓系统，除文物库房、小于5m2的卫生间及不宜用水扑救的场所外，均设置自动喷水灭火系统。博物馆内重要的展厅采用预作用系统，公共走道和办公室等部位采用常规的湿式自动喷水灭火系统。在博物馆中庭、门厅上空（净空大于12.0米）采用大空间智能型主动喷水灭火系统。系统选用智能型高空水炮灭火装置，水炮为探测器、水炮一体化设置。地下室纸质、纺织品类库房(或对空气的湿度要求很高的库房)设置全淹没气体灭火系统，介质采用IG541洁净气体。系统具有自动、手动及机械应急启动三种控制方式。在地下室设置专用的气体灭火储瓶间。

Water Supply and Drainage Design

The water the building is supplied from two water sources, specifically the water of outdoor greening and indoor flushing water is reclaimed water supplied from the municipal reclaimed water pipe, and the water of all other water consumption points is supplied from the urban water supply. For the tap water, the maximum daily water consumption is 310.2m3, the maximum hourly water consumption is 18.3m3; and for the reclaimed water, the maximum daily water consumption is 119.9m3, and the maximum hourly water consumption is 36.2m3. For the domestic water of the project, the water consumption point of the second floor and below is directly supplied with municipal pressure, and the water consumption point above the second floor is supplied by means of secondary pressurization. The basement is equipped with a domestic water tank of 60m3 and a set of water supply equipment with frequency conversion and constant pressure, wherein the domestic water tank is equipped with a secondary disinfection device. For the reclaimed water, the water is supplied by means of a water supply device without negative pressure. The indoor part is provided with a sewage and waster confluence mode and equipped with a specific ventilating pipe, and the outdoor drainage is carried out by dividing the rain and sewage. The domestic sewage will be treated through two outdoor septic tanks of 75m3, the sewage of the restaurant and kitchen will be discharged into the municipal sewer after treating by the oil separation tank. The designed return period for the roof rainwater is 10 years, and the drainage system is provided with a safety overflow port. The roof is provided with a pressure flow (siphon) drainage system, the rainwater of the platform is provided with a gravity flow drainage system to collect and drain to the outdoor inspection wells or rain ditches.

The outdoor fire water is supplied from the municipal fire hydrant. The indoor fire hydrant system is a temporary high-pressure water supply system. The fire pump of the first underground floor is equipped with a pressure water supply equipment of the fire hydrant system, wherein the pressurization of fire protection is realized with a main pump, a pressure maintaining pump and a pressure tank. The start and stop of the pressure maintaining pump and the main pump are automatically controlled through the pressure sensor on the discharge pipe. The whole building is equipped with an indoor fire hydrant system, and the water supply pipe network of the fire hydrant is circularly arranged in each floor to ensure the reach of the two substantial water columns of fire hydrant of any point. All the places (except for the cultural relic storehouses, the toilet of less than 5m2 and the place cannot be put out with water) are equipped with an automatic sprinkler system. The important exhibition hall of the museum is provided with a pre-action system, the public corridors and offices and other parts are provided with a conventional wet type automatic sprinkler system, wherein all the systems are designed according to the first level of intermediate risk, the designed strength of water spray is 6L/min.m2, and the action area is 160m2. The maximum water consumption of the automatic sprinkler system is 30L/s, the pressure required is 0.76MPa, and the water is supplied by pressurizing the main spraying pump of the fire pump room. The wet and closed spraying pressurization is carried out through a fire pool and a main pump (a pressure maintaining pump and a pressure tank). The spraying pressurization of pre-action is carried out through a fire pool, a main pump (a pressure maintaining pump and a pressure tank) and an air compressor. The courtyard and overhead of the hallway (the headroom is more than 12.0m) of the museum is equipped with an intelligent active extinguishing system, wherein the system is an intelligent high-altitude water cannon fire extinguishing device, the water cannon is an integrated device of a detector and a water cannon.The storehouses (or the storehouse with high requirements on the humidity of the air) of paper and textile in the basement is equipped with a total flooding gas extinguishing system, wherein the medium thereof is clean IG541 gas. The system is featured with three control modes, i.e. automatic, manual and mechanical emergency start. Moreover, the basement is equipped with a dedicated gas fire extinguishing bottle reservoir.

电气设计

变、配电系统的设计是本工程的核心内容。设计将本工程的信息网络、安全防范系统、消防设备、应急照明、公共区域照明、展厅用电、藏品库房用电等负荷列为一级负荷，其中信息网络、安全防范系统、应急照明为一级负荷中的特别重要负荷。采用10kV配电系统及0.4kV配电系统。照明系统的设计以及照明灯具、照明光源的选用，也是本工程的重要组成部分。在办公室、阅览室采用了高效的格栅灯盘及T5荧光灯管，在走道，休息室选用了LED筒灯，LED灯带设置于主入口及时光隧道、未来大厅等高大空间。由于采用了大量的节能灯具，大大降低了照明用电负荷。

两路10kV进线引自文化中心不同的35kV变电站10kV出线间隔，两路电源同时工作。根据当地供电部门的习惯做法，10kV配电系统不设母联。采用蓄电池作为应急照明的应急电源，UPS作为信息网络、安全防范系统的应急电源，以解决一级负荷特别重要负荷的供电要求。0.4kV低压系统为单母线分段,每两台变压器为一组，互为联络，联络开关手动控制。当其中一台变压器因故障退出运行时，切除其进线开关及其所带的三级负荷，母联开关投入，由另一台变压器承担全部一二级负荷运行，故障解除后复位，以正常方式运行。低压主进线开关与联络开关之间设电气连锁，任何情况下只能合其中的两个开关。

照明系统的设计以及照明灯具、照明光源的选用，也是本工程的重要组成部分。在办公室、阅览室采用了高效的格栅灯盘及T5荧光灯管，在走道，休息室选用了LED筒灯，LED灯带设置于主入口及时光隧道、未来大厅等高大空间。由于采用了大量的节能灯具，大大降低了照明用电负荷。

智能照明控制系统及数字调光系统的设置，满足不同时段、不同场景的控制。在达到规定照度要求、创造舒适环境的同时，取得最佳节能效果。其中数字调光系统能实现以下功能：

（1）充分利用自然采光，根据环境照度自动调整灯光亮度，无须人工参与，从而保持恒定的照度输出；

（2）在区域内安装移动探测器，做到人来灯亮，人走灯灭；

（3）普通的开关照明开灯时总保持在100％的亮灯状态，而采用数字智能照明控制系统后，可以通过预先设定的照明场景自动调整灯组的亮度。

（4）通过高效的数字控制装置（如DALI、DSI ECO镇流器等），降低镇流器的能耗，对于大量的日光灯镇流器，其节约的能源是相当可观的。

（5）系统可预先设定各种个性化的照明场景以适应不同照明环境的需求。

（6）不仅可发送控制指令，而且可反馈数字控制装置如镇流器、光源的工作状态和故障信息。

Electrical Design

The design of the power transformation and distribution system is the core content of the project. According to the importance of the building as well as national regulations and the combination of local power supply habits, the information network, safety protection system, fire-fighting equipment, emergency lighting, lighting in public areas, power consumption of exhibition hall, power consumption of storehouse and other loads of the project is listed as the load of the first grade, wherein the information network, safety protection system and emergency lighting are especially important loads of the load of the first grade. The following 10kV power distribution system and 0.4kV power distribution system are used.

The installation of the intelligent lighting control system and the digital dimming system is in line with the control of different time periods and different sites, thereby achieving the best energy-saving effect while meeting the specified illuminance requirements and creating a comfortable environment, wherein the digital dimming system can realize the following functions:

1. Make full use of natural light to automatically adjust light levels according to the ambient illuminance, in which human intervention is prevented, and the constant output of the illuminance is maintained;

2. Install a motion detector within the region to realize the purpose of turning on the lamp when there is a person and turning off the lamp when the person is leave;

3. Control the ordinary switch in 100% lighting state in case of lighting, and automatically adjust the brightness of the light group through the digital and intelligent lighting control system and the pre-set lighting scenes.

4. Reduce the energy consumption of the ballast through the efficient digital control device (such as DALI, DSI, ECO ballast and etc.), wherein the conserved energy source for a large number of fluorescent lamp ballasts is quite considerable.

5. The system can be pre-provided with a variety of personalized lighting scenes to adapt to the requirements of different lighting environments.

6. Send a control command, and feed the command to the working status and failure information of the ballast and the light source to the digital control devices to facilitate the normal maintenance and centralized management.

智能化设计

在建筑物内构筑统一通讯平台，为展览运营、物业管理以及文化交流等提供语音、数据、宽带（有线/无线）、视频、TV等等服务，创建语音、数据、视频三网融合无缝集成和有线、无线相辅相成的通讯环境。并配合绿色建筑新技术、新功能的应用和拓展，构建灵活多样的通讯技术支撑体系，配合功能完善的信息化应用体系，为建筑物内的人群提供高质量的服务。

安全防范是博物馆的重心，本项目按一级安全防范系统要求设计，选择纵深防护体系。系统采用集成式联网型安全防范系统，由视频安防监控系统、入侵报警及声音复核系统、出入口管理系统、电子巡查系统、无线对讲系统、安防信息综合管理系统等组成。

室内环境的控制主要体现文物存放环境安全、以人为本的应用需求，力求用较少的能源创造有利于文物存放、展览、使用者健康和激发使用者创新能力的室内生态环境，室内环境品质包括声、光、热等诸多环境物理因素以及室内空气品质中的诸多物理的和化学的因素，这各种因素除采用自然环境来保障外，更多依靠机械、电气等设施，创造一种扬自然环境之长、避自然环境之短的室内环境，以维持室内环境的生态平衡，而智能化系统就为创造这种环境提供技术服务。对博物馆的藏品收藏保管环境以恒温恒湿为主，分类设定不同温湿度要求进行监控；展览陈列环境以恒温恒湿与舒适性相结合进行监控；在收藏环境中，优先考虑湿度控制；对文物修复区域设置空气质量探测要求（有毒气体浓度探测），以实现人员工作环境的安全监测等。

Intelligent Design

The interior of the building is constructed with a uniform communication platform for providing voice, data, broadband (wired/wireless), video, TV and other services for the exhibition operation, property management, cultural exchange and etc., and creating a seamlessly integrated and wired and wireless communication environment with integrated voice, data and video networks, to build a flexible communications technical support system by combining with new green building technologies and the application and development of new function, and provide high-quality services for the users within the building by combining with the informatization application system with complete functions.

Safety precaution is the focus of the museum. The project is designed according to the requirements of the safety prevention system of Class I, and a depth protection system is selected therein. The system is provided with an integrated networked safety protection system composed of a video safety monitoring system, an intrusion alarm and sound review system, an exit and entrance management system, an electronic inspection system, a wireless intercom system, a safety protection comprehensive information management system and etc.

The control of indoor environment mainly reflects the safety of cultural relic store environment and the people-oriented application requirements, and strives to create an indoor ecological environment, which is beneficial to the storage and exhibition of cultural relics, the health of the user and the stimulation for the innovation capacity of user, by using less energy, wherein quality of indoor environment comprises sound, light, heat and other environmental physical factors, as well as many physical and chemical factors in the indoor air quality, these factors, protected with natural environment, rely more on mechanical, electrical and other facilities to create an indoor environment developing the advantage of the natural environment and preventing the disadvantage of the same, so as to maintain the ecological balance of the indoor environment. The intelligent system provides the technical services for creating the environment. For the storage environment of collections in the museum, constant temperature and humidity is the basis, the different temperature and humidity set according to the classification shall be monitored, the combination for the constant temperature and humidity and comfort of the exhibition environment shall be monitored. In the collection environment, the humidity control shall be considered firstly, the conservation area shall be provided with air quality detection requirements (the detection for the concentration of toxic gas) to monitor the safety of the worker.

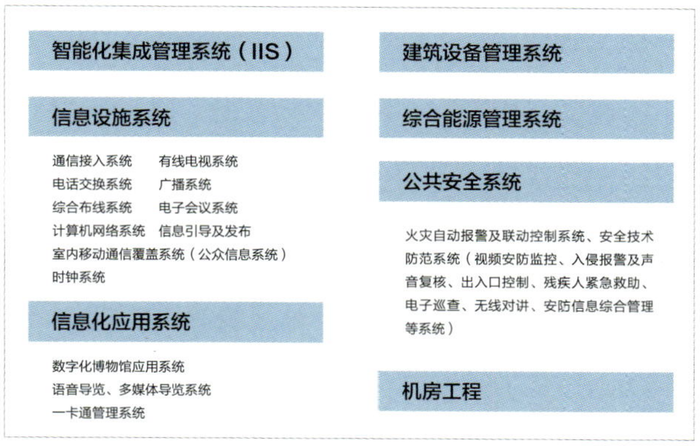

智能化系统图
Diagram of Intelligent System

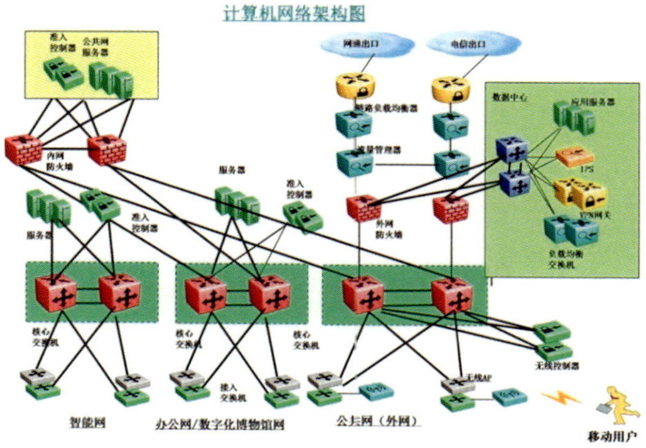

通信网络框架图
Frame Diagram of Communication Network

美术馆 Art Museum

KSP尤根· 恩格尔建筑师事务所
KSP Juergen Engel Architeken
天津市建筑设计院
Tianjin Architecture Design Institute

建筑设计

天津美术馆设计目标是建立一座集收藏、展览、研究、美育、休闲功能为一体的，以展示现当代美术艺术作品为主的大型专业美术馆，使它成为保护、积累人类文化精神和艺术财富并使之绵延传世的艺术殿堂。

作为承载艺术的建筑本身的天津美术馆也应是承载人类想象力和创造力的空间的艺术品。这一艺术品将成为天津文化和城市建设发展水平及开放形象的重要标志之一。

天津美术馆所坐落的文化中心的区域规划，旨在建立一个包含了文化及都市元素的新的城市机理。在这个区域内包含了一个大型的景观湖，区域周边是高密度的建筑组群。区域内的建筑及景观的设计环绕着自然景观的湖面展开。

天津美术馆内除了为津派美术、西方艺术、雕塑以及现代艺术提供长期的展览场所，还设有其他展厅空间可供布置巡回展览。这座坐落在湖畔的林荫大道旁的场馆，如同生长在那里一样，是一座引人注目并为之向往的地方。

Architectural Design

Tianjin Art Museum is designed to be built into a large professional gallery integrating collection, exhibition, research, art education and relaxation functions and focusing on exhibiting modern art works and to become an art gallery that protects, accumulates and spread cultural, spiritual and artistic wealth of human beings.

As a building carrying art, Tianjin Art Museum itself should be a space artwork bearing imagination and creativity of human. This art will become one of the significant symbols of Tianjin culture of and city construction development level.

Regional planning of the cultural center, where Tianjin Art Museum located in is intended to build a new urban mechanism containing culture and urban elements. There is a large landscape lake in this area and there are high density building groups around the area. Design of buildings and landscape inside is developed surrounding the landscape lake.

Besides providing long-term exhibition site for Jin style art, Western art, sculpture and modern arts, Tianjin Art Museum also has other hall space to arrange travelling exhibition. This stadium, which is on the boulevard by the lake, is an attractive and alluring place and it seems as if it is growing there.

摄影：舒赫　刘东
Photographer：Shu He, Liu Dong

北立面
North View

天津美术馆意图将人们对艺术品的欣赏不仅仅局限在建筑物内，艺术是人们来自于自然的灵感以及对美好自然的感激的表达。美术馆与景观湖之间的湖岸林荫大道扩展为一个室外的艺术展示平台，称之为“馆前雕塑广场”，参观者可经由此处进入面向湖岸美术馆向外延伸的入口回廊。入口的回廊在美术馆建筑真正的入口前形成了一个庭院，人们从自然环境进入建筑内部的体验经过了一种过渡而不至于突兀。这种独特的建筑入口形态把建筑的内部空间与外部空间紧密地联系在一起。

天津美术馆的主入口那面高达12米的玻璃幕墙，采用玻璃肋支撑体系，在水平方向上没有设置支撑构件。同时，幕墙的玻璃选用了无色的高透玻璃，即便是做到了双层中空，也不影响它的通透性，把室内和室外完全连接在一起，模糊室内外的界限，让艺术真正地从自然流入建筑之中。

Tianjin Art Museum is intended to not limit people's art appreciation activity inside buildings. Art is an expression of people's inspiration deriving from nature and gratefulness to beautiful nature. The lakeshore boulevard between the museum and the landscape lake is extended into an outdoor art show platform, which is named "Sculpture Square in Front of the Museum". Visitors can enter the entrance corridor facing lakeshore the gallery and extending outside. The entrance corridor forms a patio in front of the real entrance of the gallery. People enters the building from natural environment, with this transition, the inside experience will not be abrupt. This particular construction entrance shape combines inside space and outside space of the building together.

The 12 meter high glass curtain wall of the major entrance adopts glass rib support system without supporting elements in horizontal direction. Meanwhile, colorless and high transparent glass is used by the curtain wall. Even double glazing unit will not influence its permeability. It combines inside and outside together and makes art really flow into buildings from nature.

西南立面
Southwest View

北立面
North View

西立面
West View

共享大厅的清水混凝土柱
Fair-Faced Concrete Pillars of the Shared Space

天津美术馆入口共享空间的阶梯式楼梯向上延伸，通向各层展厅。展厅呈环状围绕布置在逐级上升的中庭平台周围，这些平台为以艺术作品为中心的展示提供了理想的条件。作为主要展览空间的综合展厅设置在二层，这一层的东西两侧设有为参观者提供信息咨询的接待区，同时还可由沿城市街道一侧的入口到达。

这是一座完全的艺术的殿堂，容纳着大大小小的展厅：一个1000平方米的综合展厅、一个约650平方米的可作为多媒体报告厅使用的灵活展厅、一个能够拥有充沛阳光的儿童活动室、一个有8米净空能够展示大型雕塑的展厅、阅览室、修复室等等。艺术品的藏品库房布置在地下室，库房设置恒温恒湿的储藏环境，总会有一天珍贵的藏品充满整个库房，并且有一天每一件藏品都能展现在市民们的眼前。

Step stairs at shared space of the gallery entrance extends upwards and leads to hall of each floor. Hall is arranged in circle surrounding the rising atrium platform. These platforms provide ideal condition for display focusing on art works. Comprehensive exhibition hall, as the principal exhibition space, is on the second floor. There are reception areas on both east and west sides of this floor. Meanwhile, there is an entrance leading to this floor on one side of the city street.

The museum is a complete palace of art that contains showrooms of all sizes, such as a comprehensive hall of 1000 sq.m., a flexible hall of about 650 sq.m., which can be used as multimedia lecture hall, a children's playroom with abundant sunlight, a exhibition room with 8 meters headroom that can display large sculpture, reading room and restoration room and so on. Storehouse is arranged in the cellar with constant temperature and humidity. One day the storehouse will be full of valuable collections and each collection can be displayed to the citizens.

三层共享空间
Three-storeyed Shared Space

美术馆北侧有着高达12米的玻璃肋幕墙与相邻的清水混凝土柱子形成天津美术馆主入口立面，这一立面通透纯粹，吸引市民进入室内的艺术殿堂。为了能够让这些柱子尽可能显得纤细挺拔，这些本来起到城中作用的柱子表现出清水混凝土本来的样子。在美术馆混凝土材质表现出细腻柔和的一面，而不是它的粗狂。为了达到这样的效果，2010年年底的时候，在文化中心各个场馆都建立了外檐的样品墙的同时，天津美术馆还额外浇筑了两棵样柱。在冬天的时候进行浇筑及最后的打磨实验后才达到了现在的效果。可能去美术馆参观的市民们不会很关注那些柱子，但事实上，这也是被希望的。因为那些柱子本应该就在那里，和这个建筑在一起和周围的环境在一起。它们了展示在这里的艺术品提供了最好的空间，而自己默默地站在那里。

There is a 12 meter glass curtain wall at the north of the gallery, which forms the main entrance elevation of the museum with adjacent fair-faced concrete pillars. This elevation is transparent and pure, which attracts citizens to access the palace in art. In order to make these pillars seems more slender and towering, they remain the original appearance of fair-faced concrete, which are originally used to bear load. Concrete material shows its softness rather than rudeness in the museum. In order to achieve this effect, exterior sample wall was established for each stadium of the culture center and meanwhile two sample pillars were cast at the end of 2010. The current effect was achieved after pouring and final polishing in winter. May be those pillars won't be noticed by citizens who visit the gallery. However, this is also expected. Because those pillars should be there together with this building and the surrounding environment. They provide the best space for work of arts here while they just stand there silently.

入口庭院
Entrance Courtyard

入口庭院夜景
Night View of Entrance Courtyard

北立面
North View

南入口
Southern entrance

东立面
East View

屋顶格栅 Roof grate

屋顶是建筑的第五立面。随着城市中高层建筑越来越高，一些较低的建筑物的屋顶便更多地暴露在人们的视线之下。原来复杂凌乱的屋顶显然不能成为破坏整个艺术中心分为的缺点，因此在设计之初，屋顶作为建筑的第五个立面被加倍地重视起来。天津美术馆本身如同石材立方体一样没有放弃对第五立面的追求，顶部的天窗提供给室内共享空间充足的采光外，其余的放置在屋顶的各种设备隐藏在了大面积的屋顶格栅之下。

Roof is the fifth elevation of architecture. With more and more mid-rise buildings rising in cities, some lower buildings will be often exposed to people's view. Obviously previous complex and messy roof shouldn't be defect destroying atmosphere of the whole art center. Therefore, at the beginning of design, as the fifth elevation of construction, roof was double valued. Tianjin Art Museum itself did not give up pursuing style of the fifth elevation just as its pursuit to style of stone cube. Skylight on the roof can not only provide sufficient sunshine to the shared space but also hide various equipments stored on the roof under roof grate.

第五立面
The Fifth Elevation

幕墙节点 Detail Curtain Wall

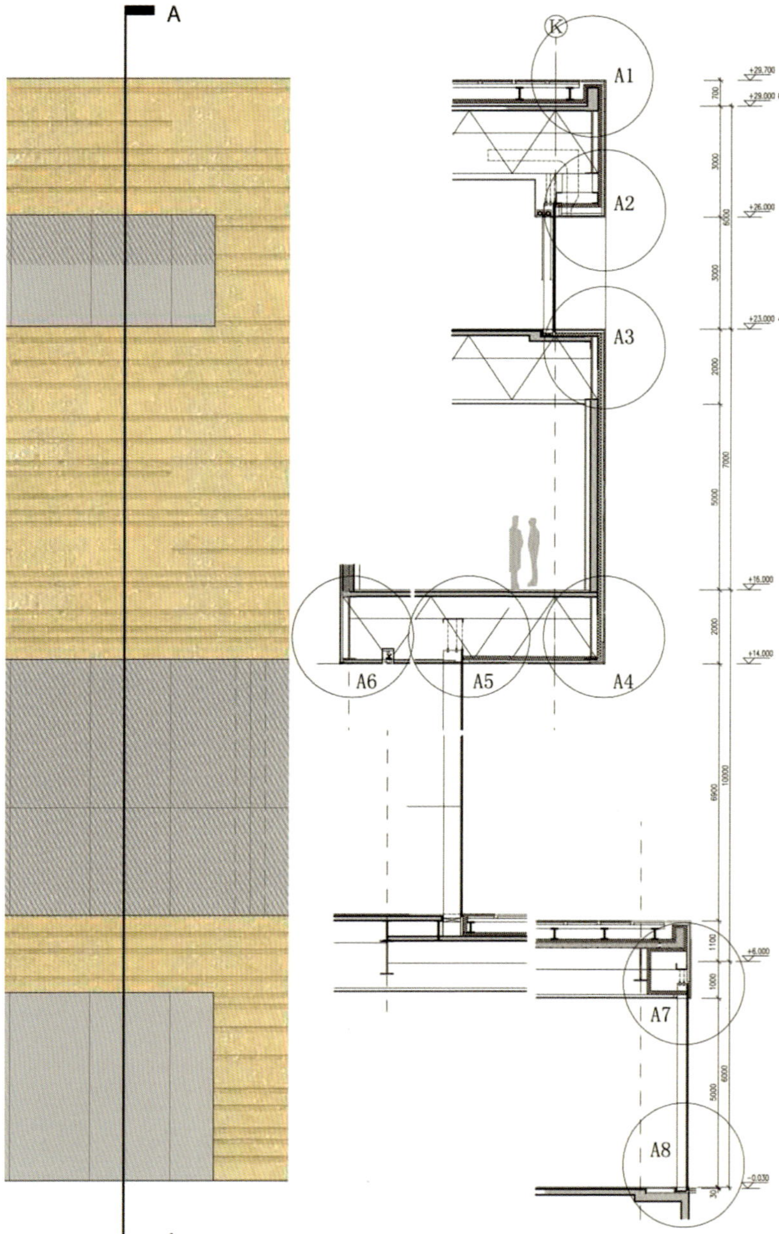

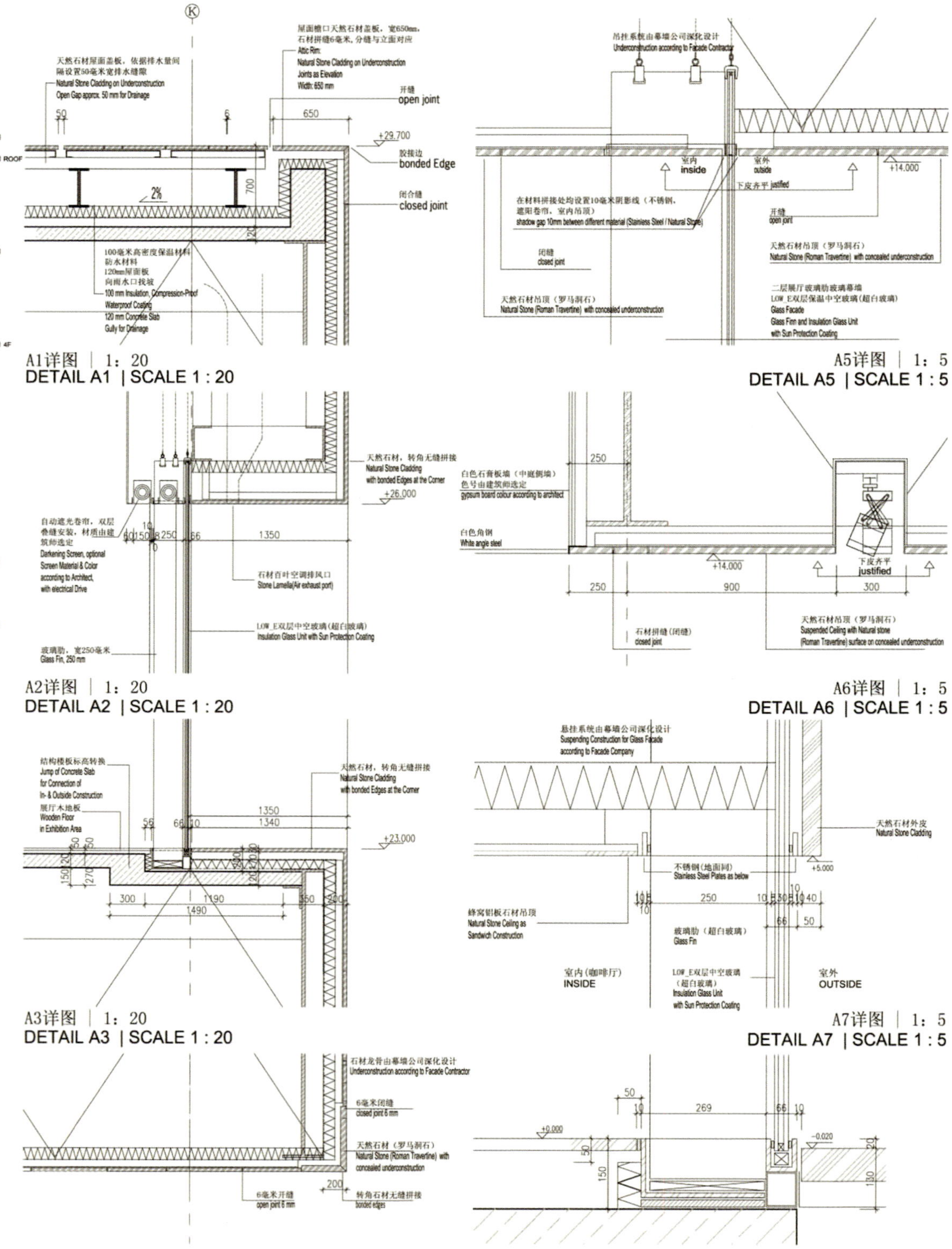

A1详图 | 1：20
DETAIL A1 | SCALE 1 : 20

A5详图 | 1：5
DETAIL A5 | SCALE 1 : 5

A2详图 | 1：20
DETAIL A2 | SCALE 1 : 20

A6详图 | 1：5
DETAIL A6 | SCALE 1 : 5

A3详图 | 1：20
DETAIL A3 | SCALE 1 : 20

A7详图 | 1：5
DETAIL A7 | SCALE 1 : 5

主入口立面外墙
The Wall of Main Entrance

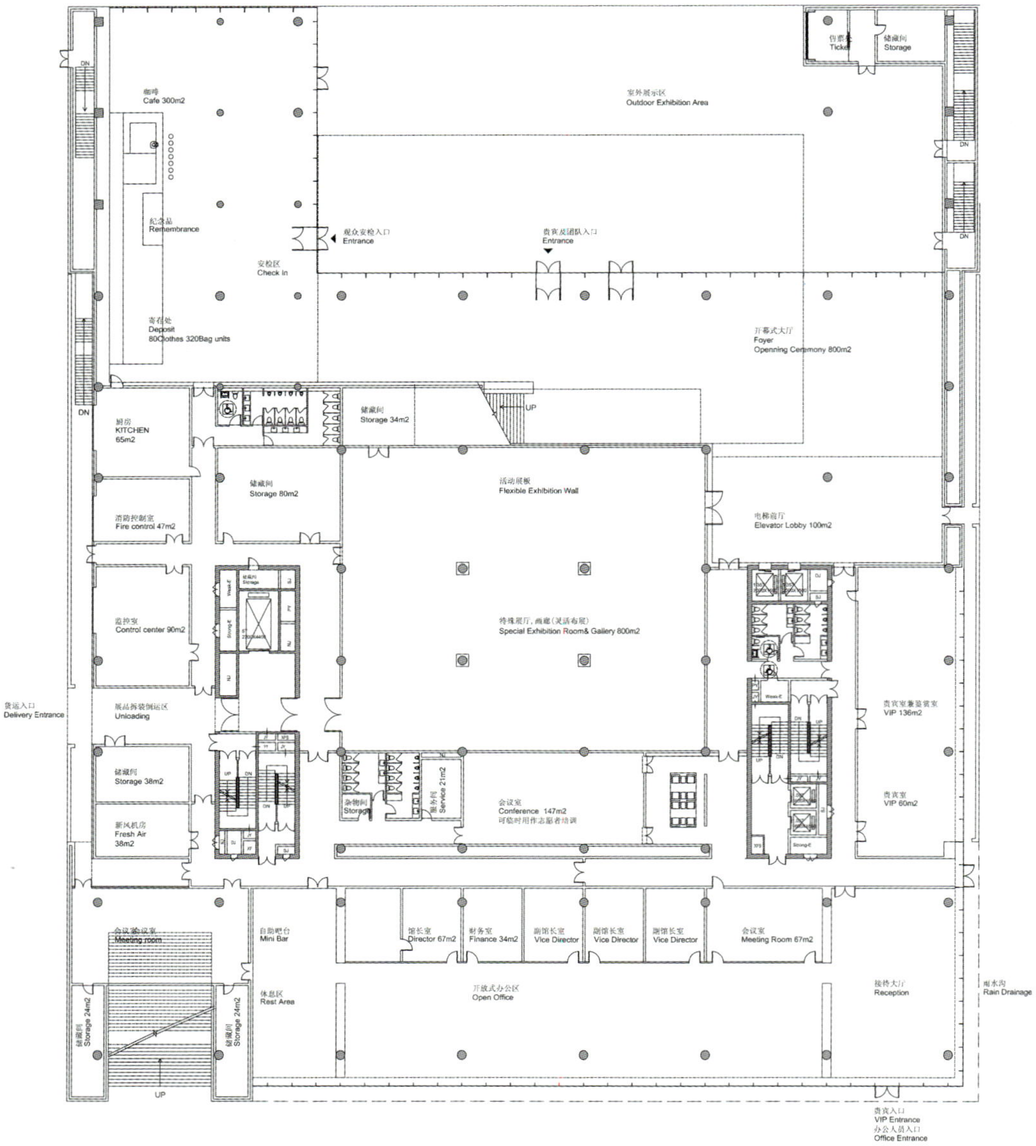

一层平面图 Floor Plan L01

二层平台
Platform of the Second Floor

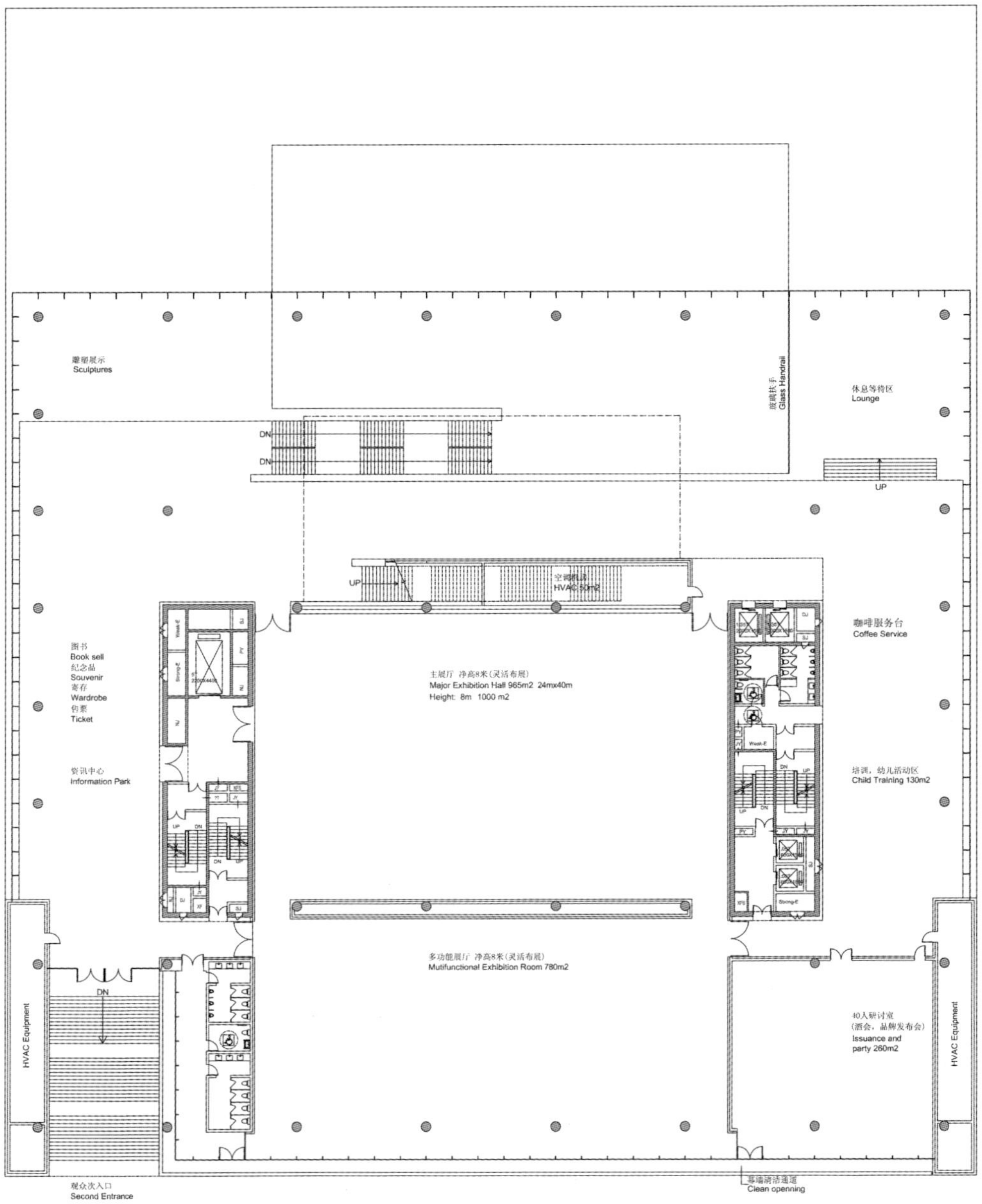

二层平面图 Floor Plan L02

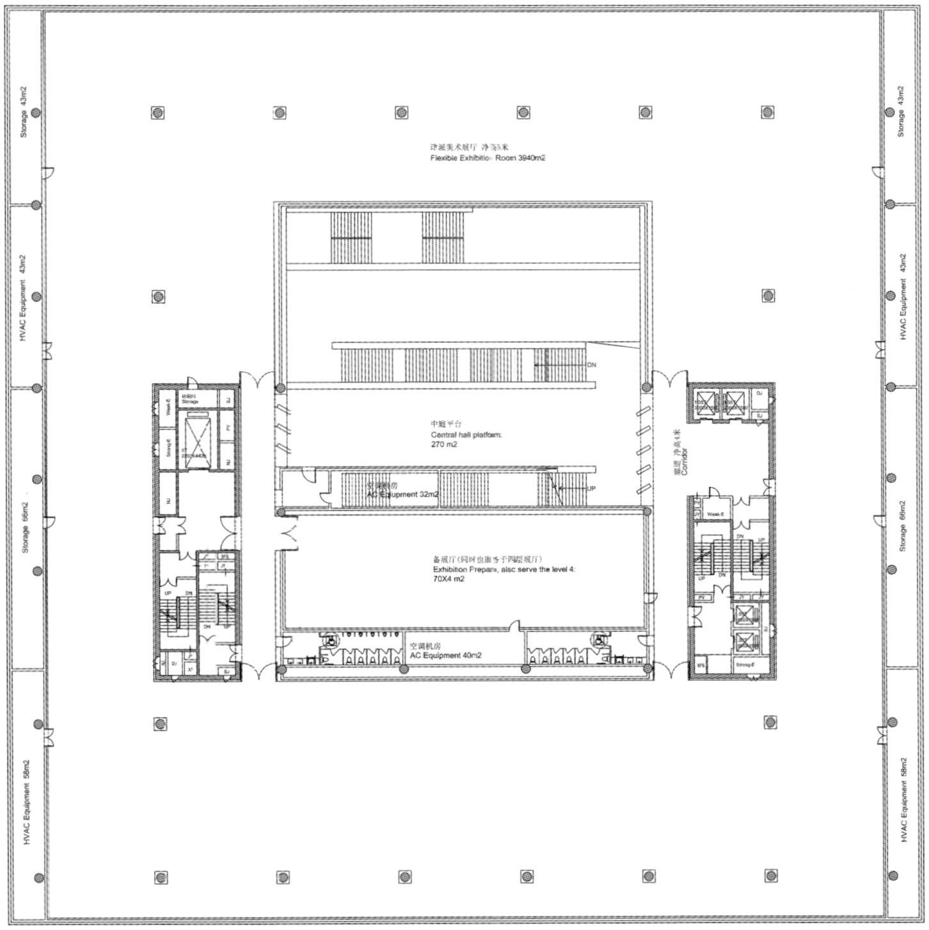

三层平面图　Floor Plan L03

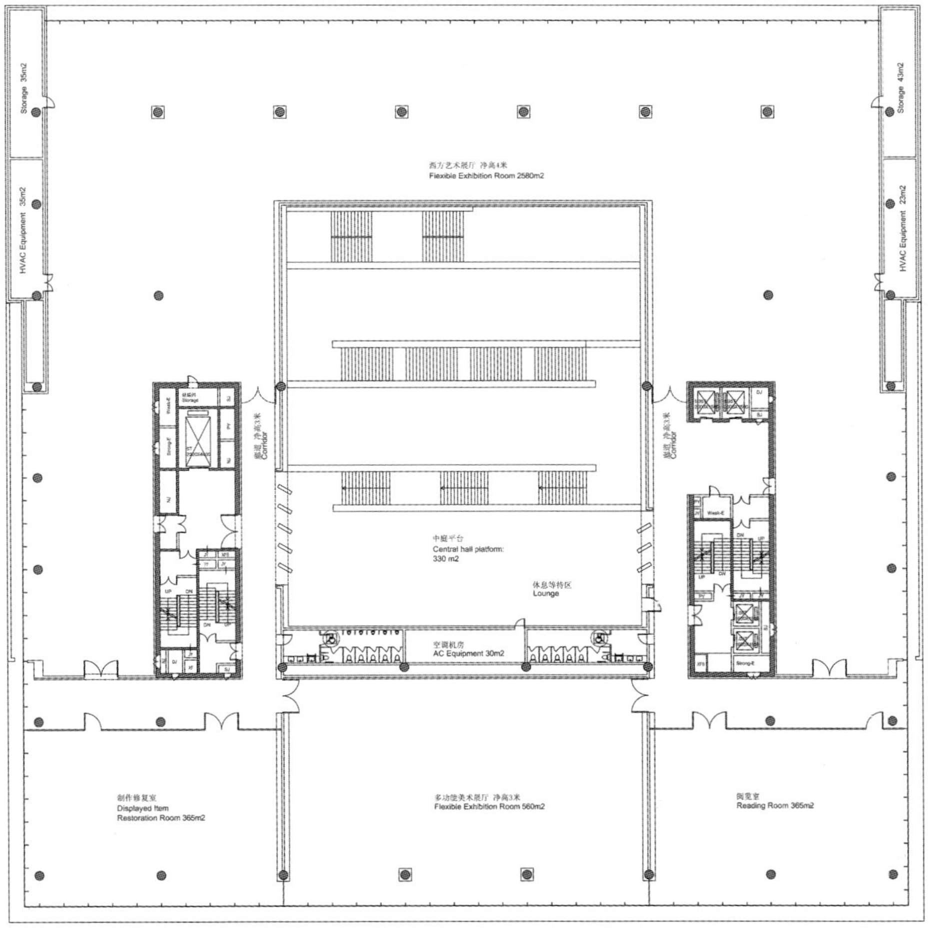

四层平面图　Floor Plan L04

首层共享大厅
First Floor of the Shared Lobby

施工过程图
Construction Process

首层共享大厅夜景
Night View of the First Floor of the Shared Lobby

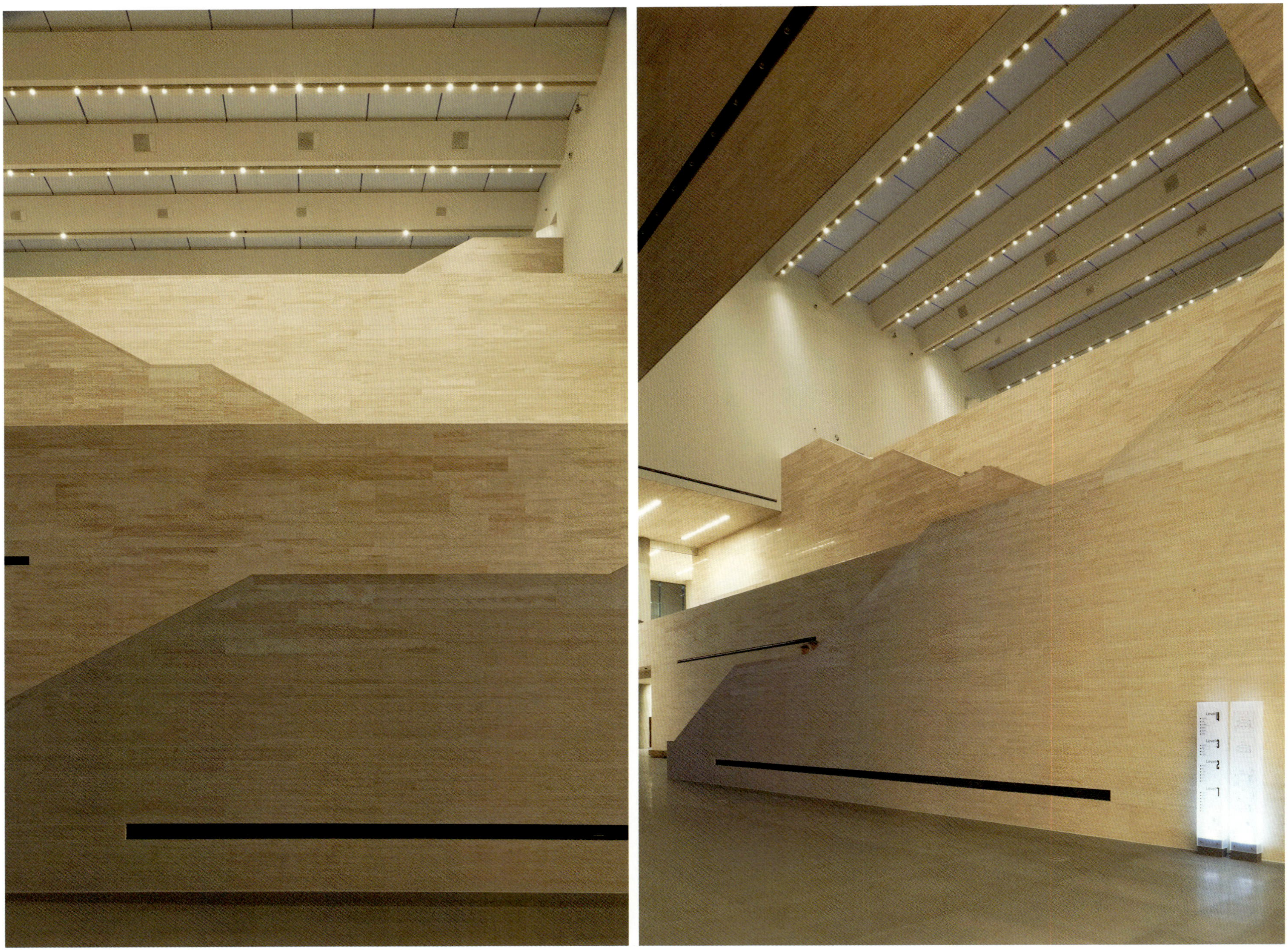

共享大厅
Shared Lobby

剖面图 Section

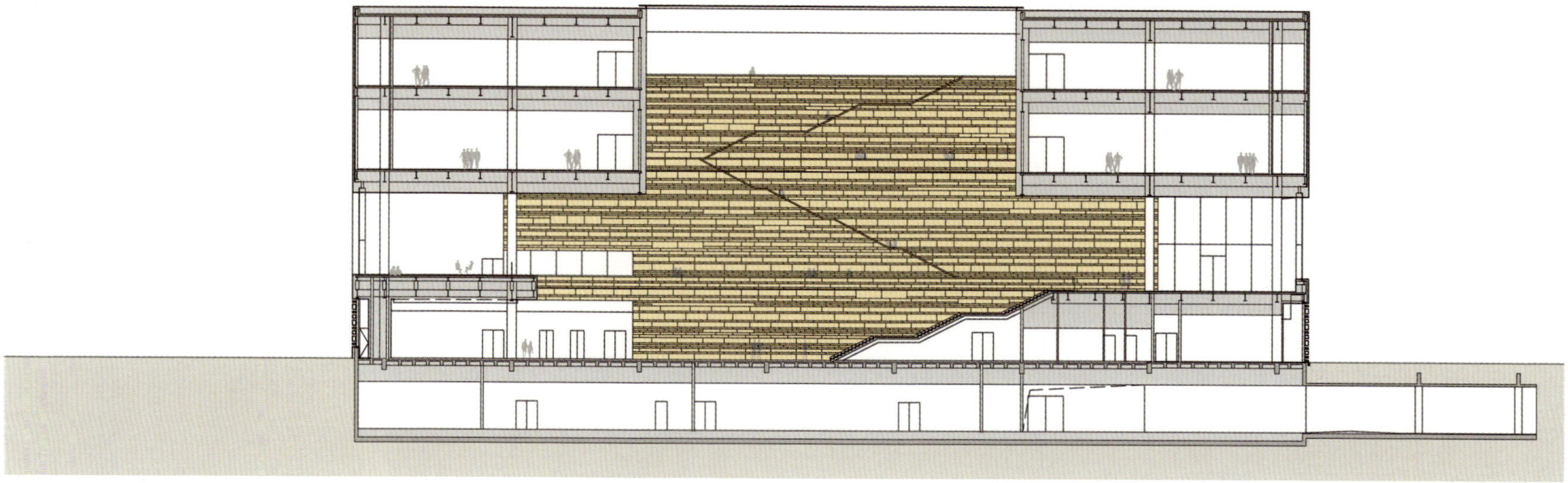

1-1 剖面图 1-1Section

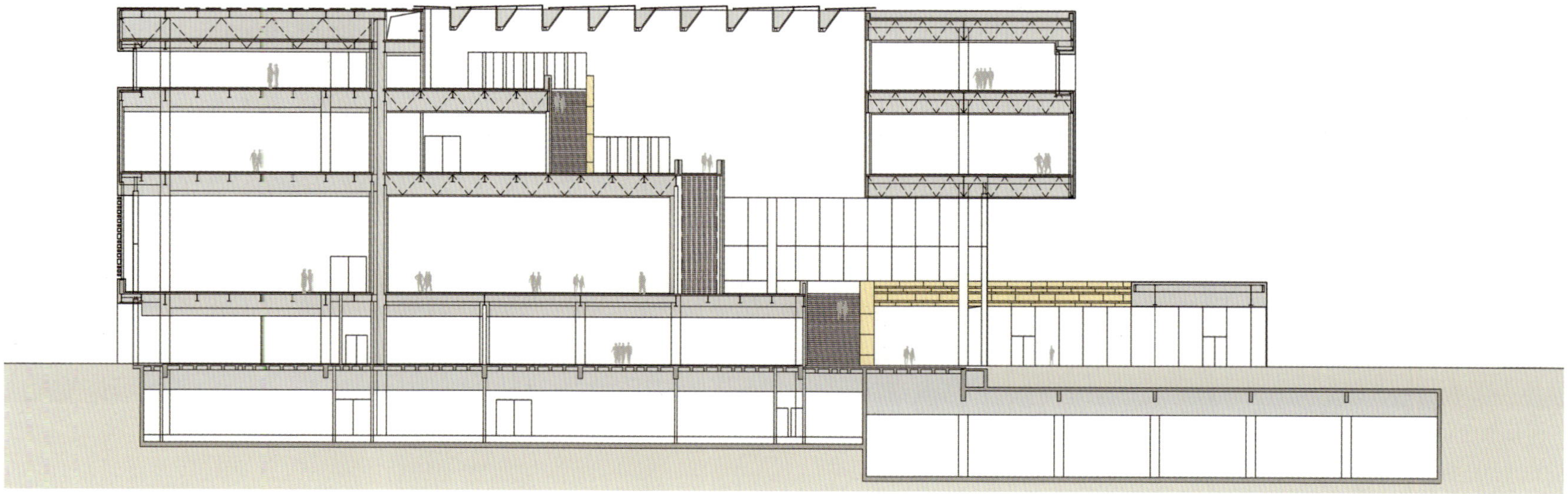

2-2 剖面图 2-2Section

二层雕塑平台
Sculpture Platform of the Second Floor

二层休息平台
Rest Platform of the Second Floor

津派展厅 Tianjin Style Exibition Hall

观景茶室 Platform Tea Room

主楼梯节点 Detail of Main Staircase

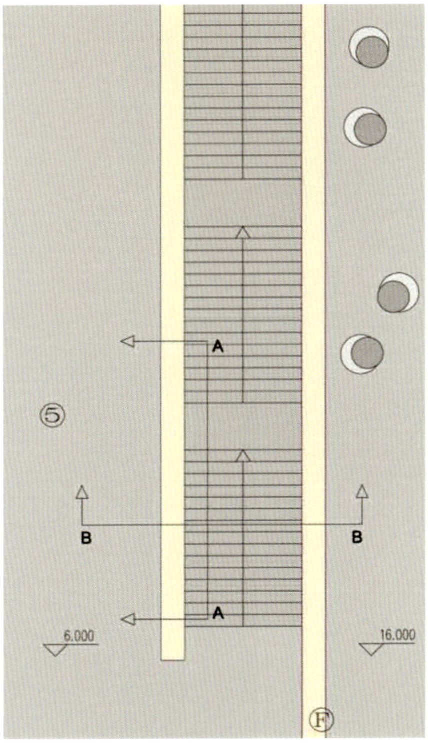

中庭二层台阶剖切图示 1:100
FLORR PLAN OF STAIRCASE 2ND FLOOR SCALE 1：100

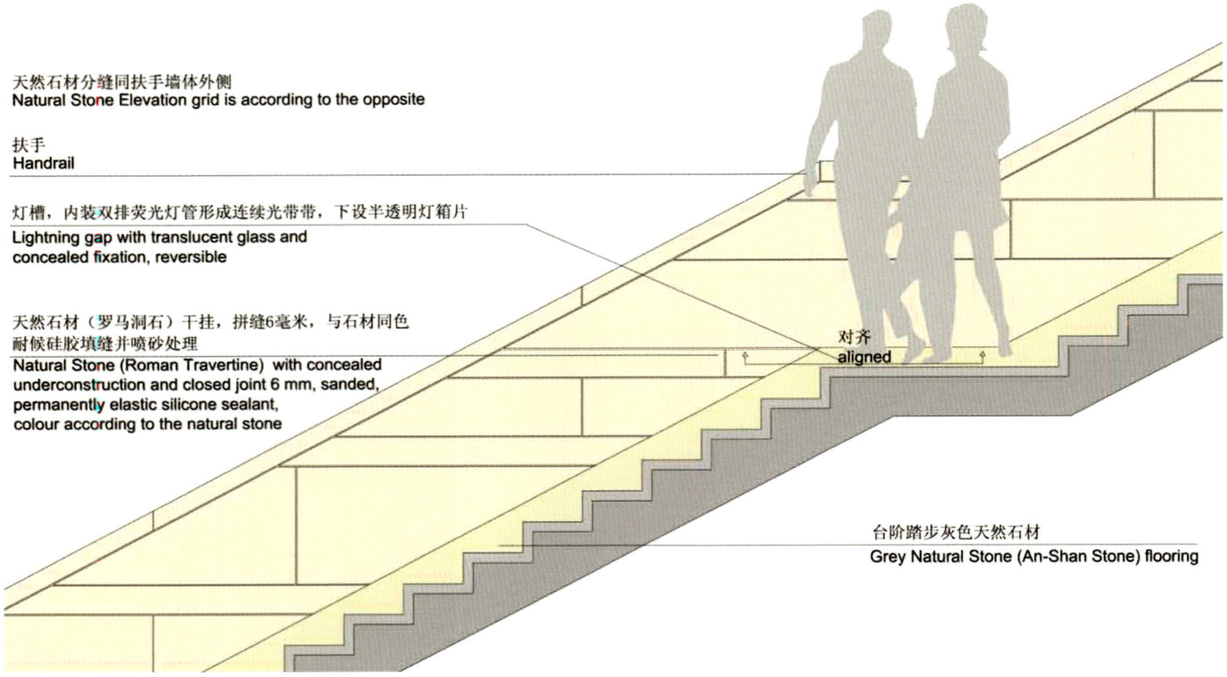

中庭二层楼梯竖向 A-A 剖切详图 1：20
VERTICAL SECTION DETAIL A-A OF STAIRCASE 2ND FLOOR SCALE 1：20

扶手及墙面天然石材（罗马洞石）示意
EXAMPLE OF NATURAL STONE RAILING (ROMANE TRAVERTINE)

地面天然灰色石材示意
EXAMPLE OF NATURAL STONE FLOORING (AN-SHAN STONE)

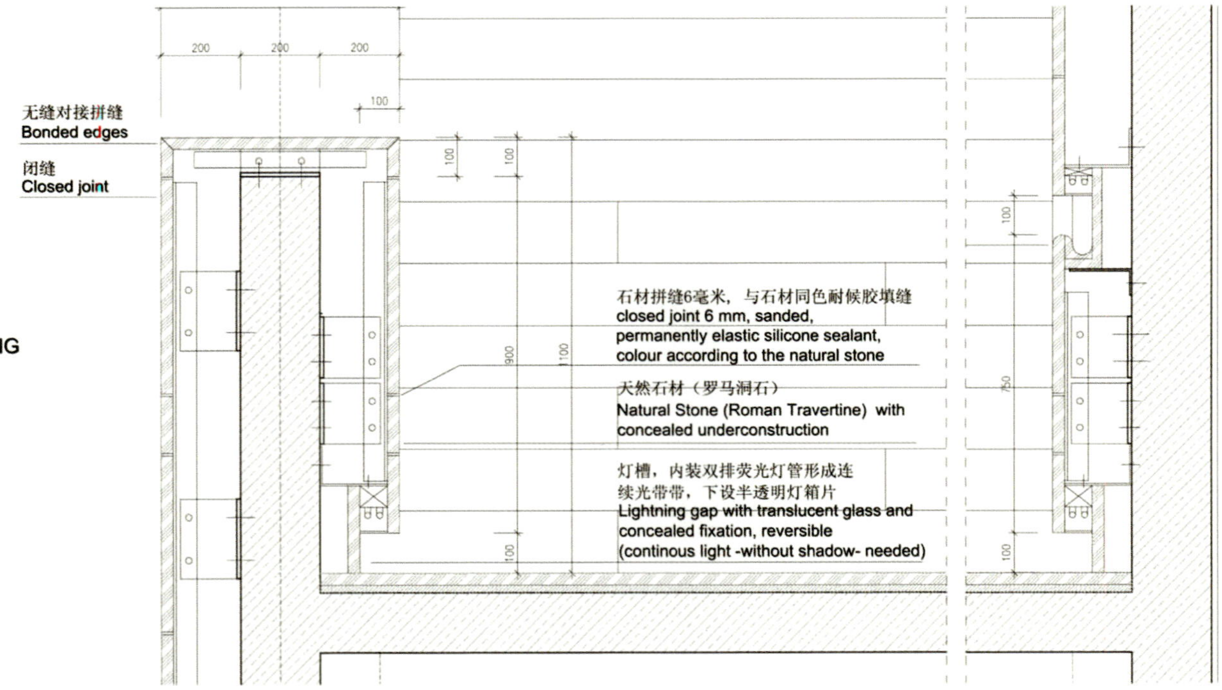

中庭二层楼梯竖向 B-B 剖切详图 1：10
VERTICAL SECTION DETAIL B-B OF STAIRCASE 2ND FLOOR SCALE 1：10

共享楼梯通往三层展厅
Shared Stairs Leading
to the Third Floor Hall

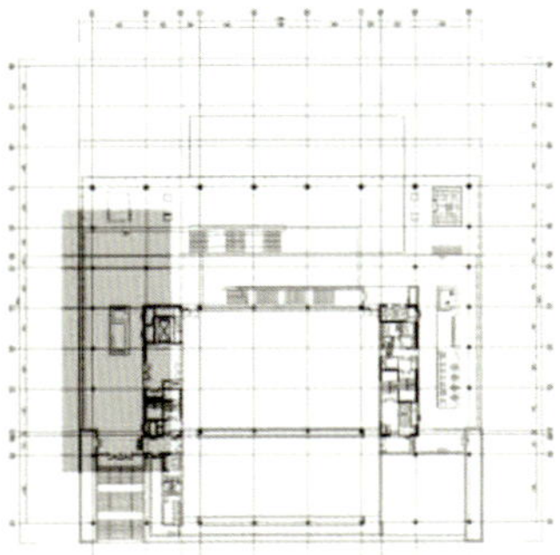

二层平面 LEVEL 2

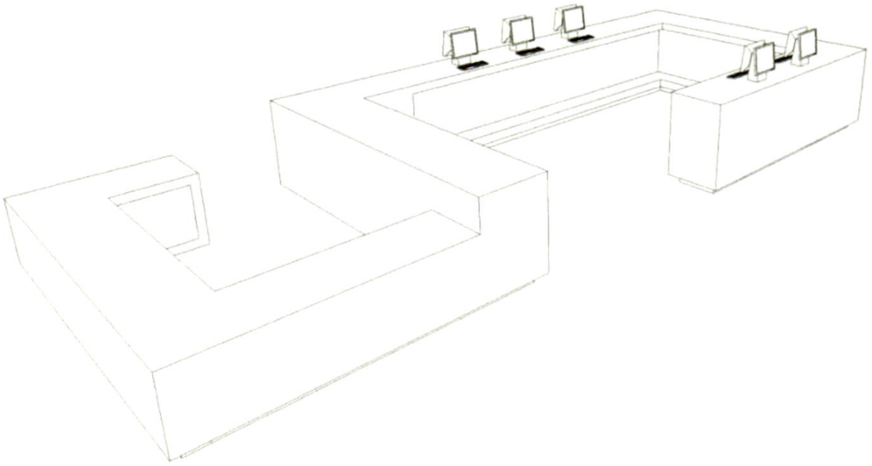

3D模型 3D VIEW

材料 MATERIAL

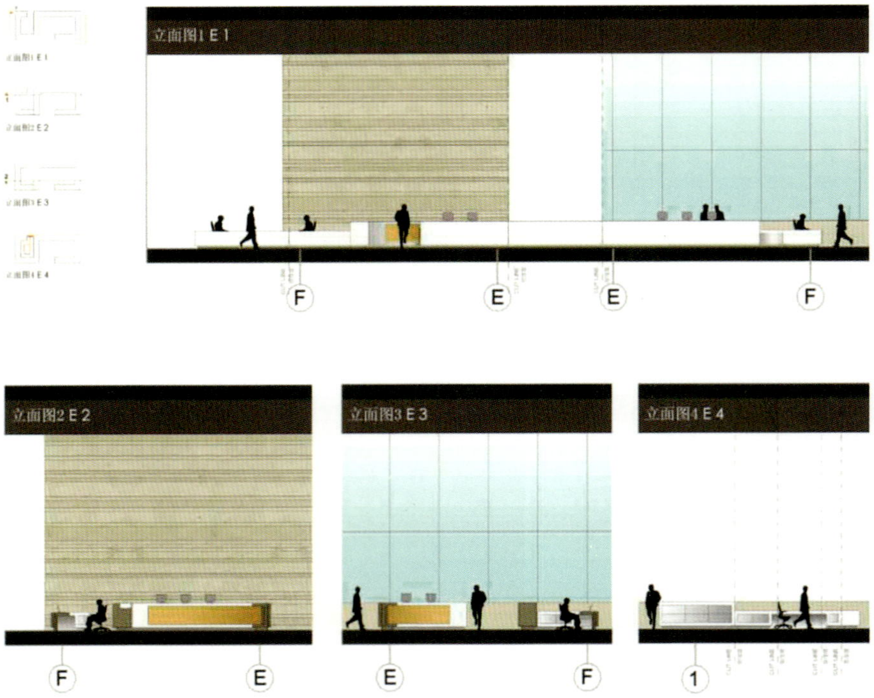

立面图 ELEVATIONS

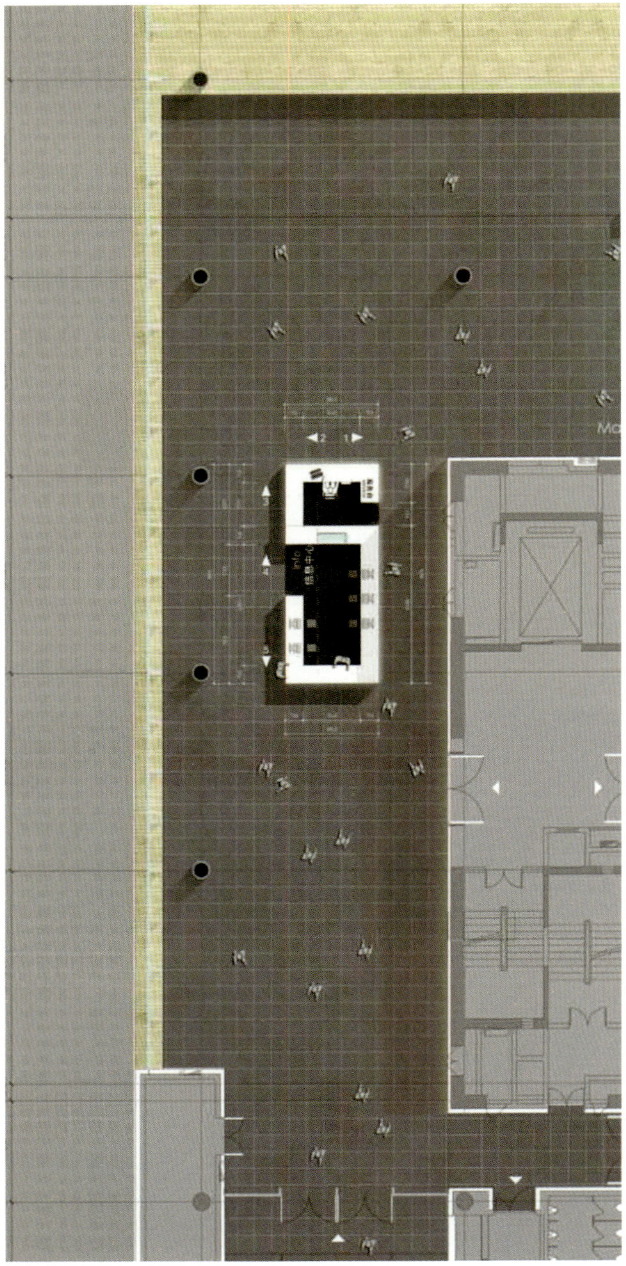

平面图 FLOORPLAN

07 _ 信息中心 INFORMATION PARK

接待台 Information Desk

西立面
West View

结构设计

本工程高29.7m，地下一层，地上四层，采用钢框架-剪力墙结构体系。首层北侧入口处大跨度桁架跨度47.1m，北侧二层至屋顶层结构悬挑楼面悬挑长度9.2m。主要柱网尺寸为10.8m×8.1m、18.3m×10.8m，内部大空间部位柱距为32.4m。结构中采用了大量桁架以实现设备管线的通过及室内净高要求，部分位置采用钢骨混凝土结构增大柱子的刚度和承载能力。

通过多角度对比分析几种形式的钢结构组合楼板在力学性能、施工周期、综合经济效益等方面的差异，提出采用经济效益更为显著的钢筋桁架模板组合楼板。采用钢结构组合楼面梁，有效地减少了用钢量，且梁高较小，利于设备专业的管线布置。对大悬挑及大跨度部分采用钢桁架结构，充分利用吊顶空间并进行设备管线综合，保证建筑方案吊顶高度的要求，实现工程的高品质需求。

主入口处大悬挑处采用内置圆钢管的劲性混凝土柱，表观效果为清水混凝土。利用设备用房及交通核设置了两个混凝土筒体，提高了整体结构的抗侧刚度和扭转刚度。对主入口14m高的全玻璃幕墙进行详细分析计算，包括节点的设计及分析，确保其安全可靠。缜密的程序分析，采用ETABS、SAP2000、PKPM、ANSYS等程序对结构进行了多遇地震下的弹性静力、弹性时程、罕遇地震下的弹塑性静力推覆等分析计算，确保结构经济、安全、高效。

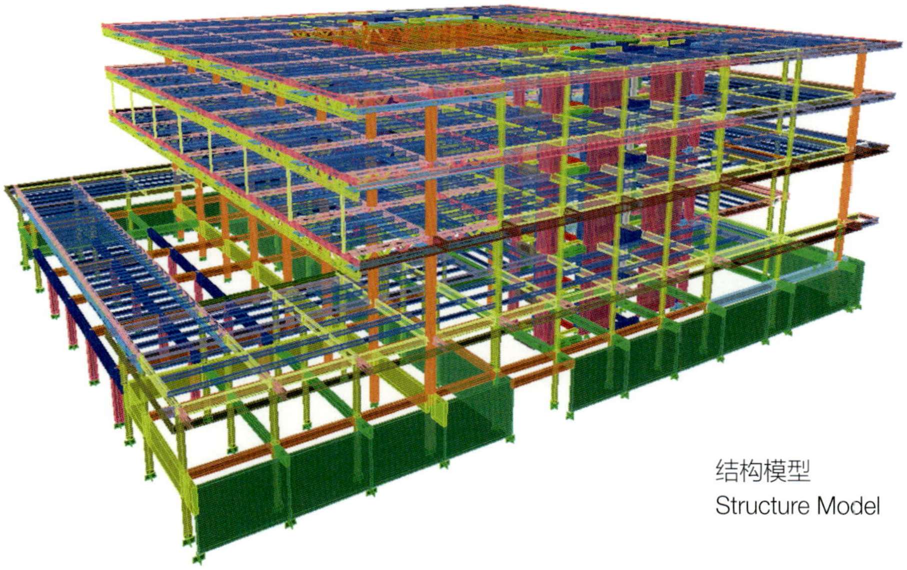

结构模型
Structure Model

幕墙施工图 Curtain Wall Construction Site

Structural Design

This project is 29.7m high with one floor underground and four floors above the ground. It adopts steel structure and shear wall structure. Span of the first floor large-span truss at the north entrance is 47.1m and overhang from the northern second floor to cantilever floor of roof structural layer is 9.2m. Dimension of major column grid is 10.8m X8.1m and 18.3m X10.8m. Column spacing of inside large space part is 32.4m. Many trusses are used to realize routing of equipment pipeline and requirements of ceiling height. Some position adopts steel reinforced concrete structure to increase stiffness and carrying capacity of columns.

Propose to adopt steel truss formwork composite deck of more significant economic benefits via comparing and analyzing differences of several steel structure composite floors in mechanical properties, construction cycle and comprehensive economic benefits from a variety of perspectives. Adoption of steel structure composite floor beam effectively reduces steel volumes and small beam depth is beneficial to professional layout of equipment pipelines. Steel truss is used for large cantilever and large-span part, ceiling space is fully used and equipment pipelines are integrated, which guarantees ceiling height requirement of construction scheme and meets high quality demand of engineering. Large cantilever at the main entrance uses composite steel concrete column with built-in circular steel tube, appearance effect of which is fair-faced concrete. Two concrete cylinders are installed by using equipment rooms and circulation hub, which enhanced lateral stiffness and torsional stiffness of the overall structure. Conduct detailed analysis and calculation to 14 meter high glass curtain wall at the main entrance, including design and analysis to connections, to ensure its security and reliability.Rigorous analysis. It analyze and calculate elastic static force and elastic time usually occurred in an earthquake and elastoplastic static pushover seldom occurred in an earthquake with ETABS,SAP2000,PKPM and ANSYS, etc. to ensure safe, cost-effective and highly efficient structure.

玻璃肋幕墙
Glass Curtain Wall

暖通空调设计

（1）高大展厅空调风系统形式与气流组织有效性

高大展厅空调系统服务对象较为特殊，不仅需要满足室内人员舒适要求，而且还必须满足室内舒适度的均匀性要求，因此为展览建筑以室内空气特性分布指标ADPI≥80%为判据，对展厅空调气流组织进行优化分析。结果表明二次回风空调系统不仅满足室内舒适度、均匀性要求，而且可实现空调系统的节能。

（2）序厅共享空间气流组织分析

本项目的步进式序厅共享空间，总高度约30m，形状特点如图3所示，为倒阶梯型。该空间内分布了四处位于不同标高的人员活动区域，其中4层共享空间人员活动区接近屋顶天窗。这一特点对共享空间的空调气流组织形式的确定提出了挑战，因为在这样的空间内，无法形成涵盖所有人员活动区的分层空调，只能采用全室空调或分别服务不同标高人员活动区的局部空调。显然，如果能形成有效的局部空调，会比采用全室空调有明显的节能效果。此问题无法采用常规工程计算解决，必须通过CFD模拟计算。模拟结果表明，局部空调可解决序庭共享空间的热舒适问题。

HVAC Design

(1) Effectiveness of air conditioning duct system type and air distribution of HVAC in tall hall

Service object of air-conditioning system in tall hall is special. It is required to not only meet comfort requirement of people indoor but to meet uniformity requirement of indoor comfort level. Therefore, optimization analysis is performed to air distribution of HVAC according to indoor air diffusion performance index of exhibition building, i.e. ADPI≥8. The results show that secondary return air system cannot only meet indoor comfort and uniformity requirements but also realize energy conservation of air-conditioning system.

(2) Analysis of air distribution in shared space of lobby

Total height of the lobby's stepwise shared space in this project is about 30m. As shown in figure 3, its shape is like an inverted ladder. This space is divided into four activity areas for people at different elevations. Amongst activity area in the fourth floor shared space is close to the roof skylight, which brings challenge to determination of HVAC air distribution in the shared space. On account that stratified air conditioning that covers all activity areas cannot be realized in such space, therefore, on full room air conditioning or partial air conditioning serving activity areas of different elevations can be adopted. It is obvious that if effective partial air conditioning can be designed, it will have more evident energy-saving effect than full room air conditioning. This problem cannot be solved by conventional engineering calculation, but must be resolved by CFD analog calculation. Partial air conditioning can meet thermal comfort requirement of the shared space.

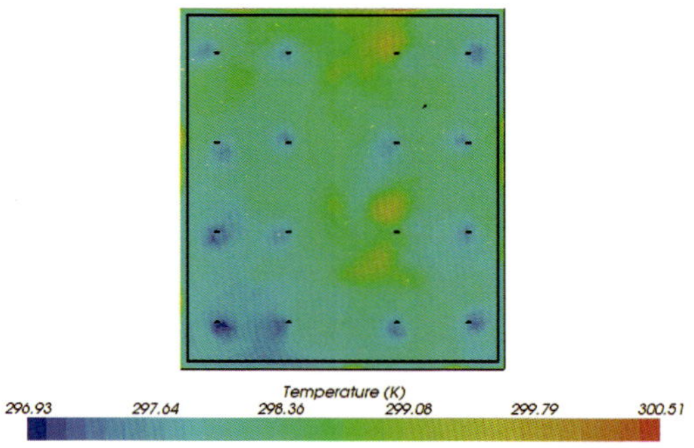

图 1 展厅 1.5m 处温度场（1.8 次换气）
Figure1. Thermal Analysis of Exibition Hall at 1.5m

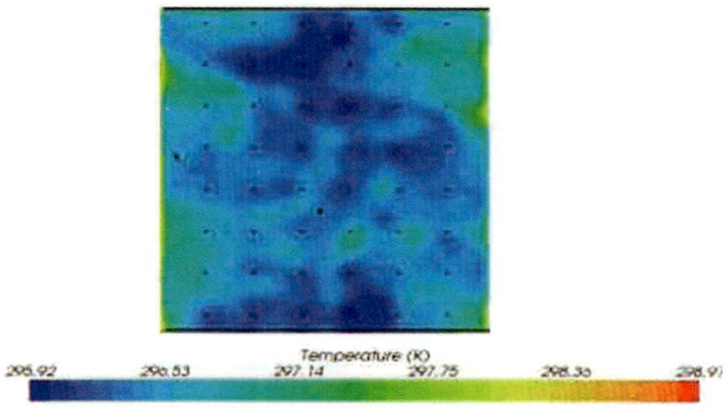

图 2 展厅 1.5m 处温度场（5 次换气）
Figure2. Thermal Analysis of Exibition Hall at 1.5m

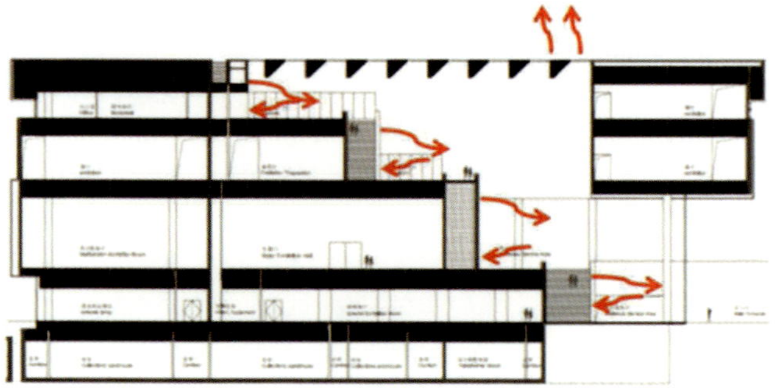

图 3 共享空间局部空调气流组织示意图
Figure3. HVAC Air Distribution of Shared Space

图 4 各层 共享空间温度分布图
Figure4. Temperture Distribution of Each Floor

给排水设计

（1）雨水系统设计

美术馆顶层设置有大面积的展厅，且由于展厅平面布置要求，雨水悬吊管无法完全避开展厅，屋面雨水系统的设计安全显得尤为重要，给设计带来了很大难度。

解决方案：

A 提高雨水系统排水能力设计标准，除设置雨水管道系统外，同时设置雨水管道溢流系统，总排水能力之和能够满足100年重现期的雨水量。

B 选用连接可靠、耐久的雨水管材，并采取防结露措施。雨水系统管材选用虹吸雨水专用高密度HDPE管材，电热熔连接，吊顶内管线均设置橡塑保温。

C 展厅区域吊顶内的雨水管线设置漏水报警装置，并纳入整个弱电系统，出现隐患能够及时向总控室报警。

（2）消防系统设计

美术馆存在大面积的展厅及展品库房，消防系统设计除应满足消防的防护功能外，同时应避免系统平时及火灾扑救过程中对展品可能造成的水渍损害。并且美术馆平面布局较为复杂，在中庭共享空间存在着不同标高的交通平台。以上都为消防系统设计带来了难度。

解决方案：

A 对于展品库房等无人封闭区域设置气体灭火系统；

B 对于展厅等大空间，且存在大量人流的区域设置预作用喷淋系统；

C 在中庭共享空间，设置自动扫描射水高空水炮系统。每个平台布置两个高空水炮，并布置在侧面，既保证了消防防护功能，又满足了室内装修设计的要求。

Water Supply and Drainage Design

（1）Design of rain water system

There is a large hall at the top floor of the gallery. On account that rain water hanger cannot completely avoid the hall to meet requirement of hall plane layout, therefore, design of roof rain water system appears to be particularly important and great difficulty was brought to the design.

Resolutions:

A. Enhance design standard about drainage capacity of rain water system. Besides water rain pipeline system, rain water overflow system is also established. Sum of total drainability can deal with rainfall of 100-year return period.

B. Select rainwater pipe of reliable connection and endurance and take condensation prevention measures. Rain water system adopts siphonic HDPE pipes specially designed for rain water and electric hot-melt connection. Rubber insulation is installed for pipelines inside the ceiling.

C Rainwater pipeline inside the ceiling of the exhibition hall is equipped with leakage alarm device that is integrated into the entire weak electricity system. If hazards appear, the mater control room can be informed promptly.

（2）Design of fire fighting system

There are large halls and storerooms for exhibits in the gallery. Besides fire protection function, meanwhile design of fire fighting system should avert possible water damage to exhibits in daily storage or exhibition and during fire suppression process. Plane layout of the gallery is complicated and there are circulation platform of different elevations in shared space of the atrium. All of the above-mentioned problems bring difficulty to design of fire fighting system.

Resolutions:

A. Gas fire extinguishing systems are installed in closed area without people, e.g. storehouse;

B. pre-action sprinkling systems are installed in large space, e.g. exhibition halls and in areas with high flow of visitors;

C. Sky fire gun system of auto-scanning jetting is installed in shared space of the atrium . Two Sky fire gun systems are installed on side of each platform, which not only ensures fire fighting function, but also meet requirements of interior design.

火泵室　Fire pump room

钢瓶室　Steel bottle room

电气设计

美术馆的照明及其智能化控制系统设计不仅实现了节能、更重要的是满足了布展需求 —— 保护展品、突出展品、还能适应不同布展布局及不同类展品布展的要求；

采取综合措施保证展品的安全，如：展厅的漏水检测系统、完善的安防立体防护体系、多点温湿度监测及恒温恒湿的空调控制系统；

智能化程度高为美术馆的现代化管理及服务奠定了基础。

Electrical Design

Lighting and Intelligent control system of Art Gallery not only realize energy saving, but also satisfy the layout requirement—protect the exhibits, highlight the exhibits, and adapt to the requirements of different layout and the layout of different type of exhibits.

Adopt comprehensive measures to ensure the safety of the exhibits, such as: the leak detection system of the exhibition hall , perfect dimensional security and protection system, multipoint temperature and humidity monitoring and constant temperature and humidity of the air conditioning control system;

The high degree of intelligence lay the foundation for modern management and service of Art Gallery.

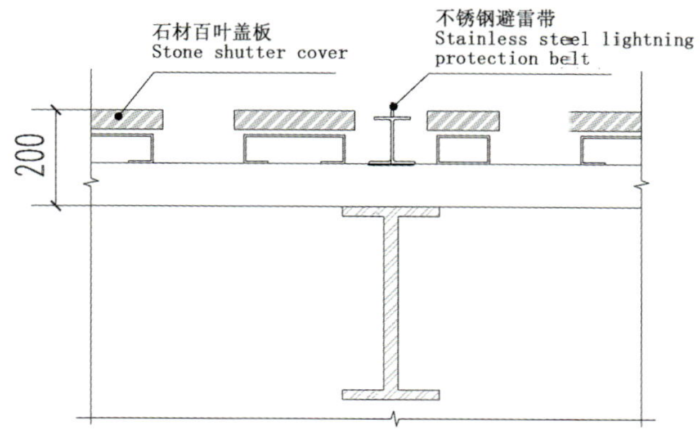

屋顶避雷带做法
Installation of Roof Lightening Protection Belt

多功能厅 Multifunctional Hall

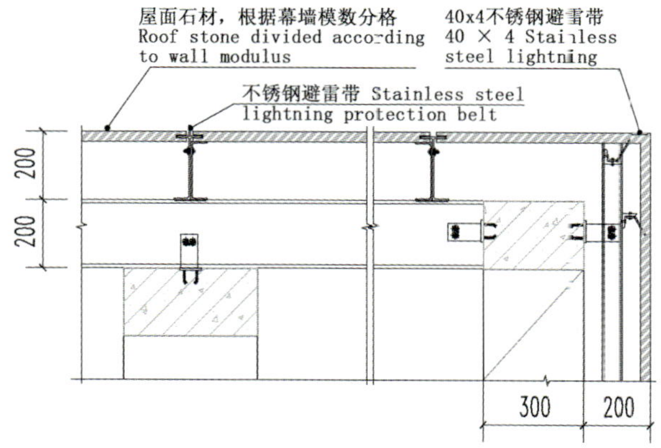

屋顶转角避雷带做法
Installation of Lightening Protection Belt at Roof Corner

图书馆 Library

株式会社山本理显设计工场
Riken Yamamoto & Field Shop
天津市城市规划设计研究院
Tianjin Urban Planning & Design Institute

建筑设计

天津图书馆藏书约600万册，除了具备一般的借阅功能，还设有展厅、报告厅、培训室、餐厅、咖啡厅等公共功能空间，以及专业阅览室、古籍修复保存等空间，具有多元化功能特点。在设计如此大型的图书馆时，目标是营造一个立体城市空间般的图书馆，一座人们热爱的知识主题乐园。让公众可以自由造访，身在其中，如同置身于街道、公园，可以轻松随意地开展多种活动。

Architectural Design

The Library's collections include around 6 million books. Public spaces in library are functional diversification such as professional reading room and room for conservation of old and rare books. Other functional spaces besides the basic borrowing function are the exhibition room, lecture room, training room, dining hall, café and etc. The initial design idea to create a library of such scale is making a three-dimensional city library and a paradise of knowledge. By making the library freely available to the public, street life atmosphere is thus created where people can hang out and relax, and linger with fun and get active.

摄影：魏刚
Photographer：Wei Gang

立体城市空间
Three Dimensional Urban Space

本展室 Rare Books Showroom 近代文献暨地方文献阅览区
nt Information 检索室 Retrieval Service 数字资源服务区
se Newspapers 报告厅 Lecture Hall
Visual Impaired Reader's Service 自习室 Study Room

5 Reserved Collections Foreign Books
4 会议厅 Conference Hall Multimedia demonstration Room
3 Back Issue of Chinese Periodicals Chinese Books
2 Chinese Periodicals Chinese Books
1 餐厅 Restaurant Library Shop 咖啡厅 Cafe
General Information Desk

三富变化的墙面
Changeful of Wall

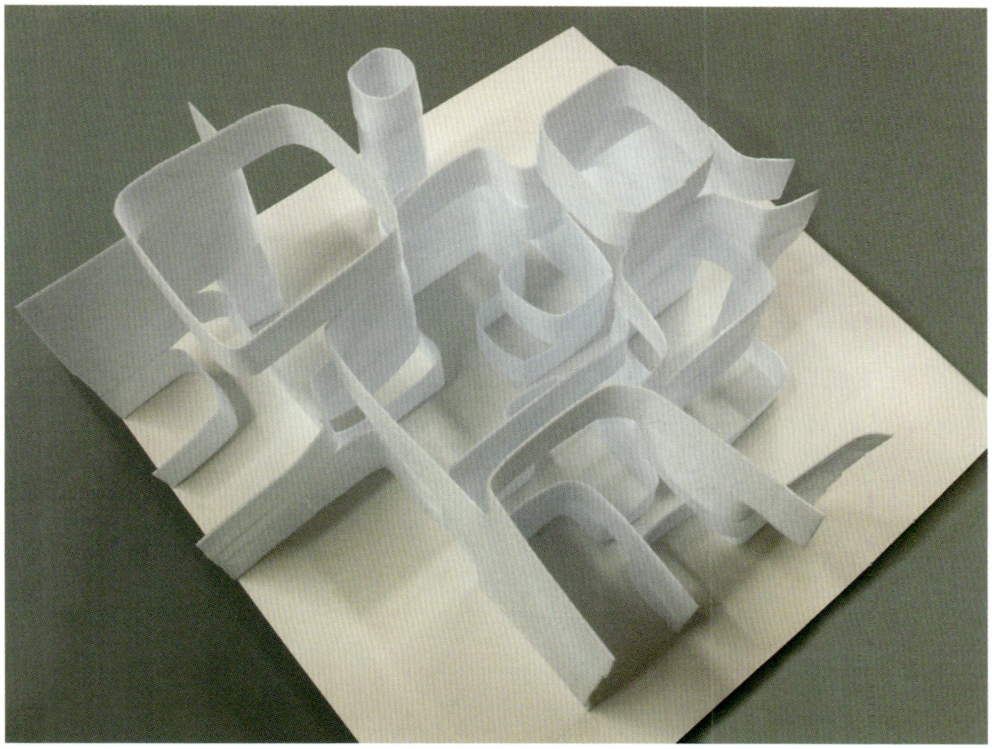

墙的概念
Wall Concept

为了实现上述空间构想，设计最初的意向就是突出墙面。有些墙面围合封闭，有些墙面开放，有些位置的墙面错位、叠和，有些墙面上布满书架，构成了各式各样的特色迥异的空间，容纳各种使用功能。这些设想从最初的意向，一步步推进坚持，一直延续到深化设计、建造阶段，直到最终完成。

In order to achieve a concept of space, the initial design idea is giving the top priority to the wall. The change of wall makes different spaces: some are enclosed spaces and some are open. Several walls run up and down, some are in different position on different floors. Some walls are stocked full of books. Spaces are given distinctive characters by the different arrangements of walls. The initial idea has been conceived from the initial concept to the accomplishment, even though having occurred all sorts of obstacles in the detailed design phase and construction phase.

墙面的内部结构采用交错桁架结构体系。高度大约6米的墙面呈20.4米的网格布置，上下层网格之间错位10.2米，相互叠合。墙面原则上沿网格布置，与功能及空间相对应，有的部分去除一段、有的部分增加一段。根据这种空间设想，柱子作为定义空间的元素完全消失，整个建筑内部形成了几乎没有柱子的大空间。其中，既布置有较为封闭的小空间，也有开放的共享空间。这些空间自然流畅的组合成一个整体。人们漫步其间，犹如在街道中闲游般，自在地体验各处的景致。

各层的功能构成非常简洁易懂。首层以南北贯穿的通廊式大厅为中心，布置了总接待台、展厅、自习室、书店、餐厅和咖啡厅。各种公共功能集合在一起，便于读者使用。越向上层走，图书的专业性越强。二～三层是一般图书和杂志、报纸阅览室。四层是电子阅览室、视听阅览室。五层则是专业阅览室及古籍修复、保存等空间。

The inner structure of wall uses the girder truss structural system and gridding arrangement, which is 6 meters high. The whole structure is set along the 20.4-meter grid horizontally and vertically. Each structural squares stagger and is composite of 10.2-meter. Walls are add up in different part of the structure. With the concept of partitions and wall dividing up library spaces, columns are canceled which opens up a big room in the library. Enclosed small space and open sharing space are in this big room to form an integrated space. Hang around in the library creates a leisurely and relaxed street walking feel while appreciating surrounding scenery.

Function composition is easy to use. First floor is centered around a corridor running through south to north. Reception desk, exhibition hall, individual study room, book store, dining hall and café are set in the corridor in aims of convenient use for readers. Book inventories are gradually more professional while walking upstairs. General books, magazines and newspaper reading room are set at the second and third floor while digital reading room and video and audio room are on the fourth floor. Professional reading room and room for conservation of old and rare books are set at the highest level on the fifth floor.

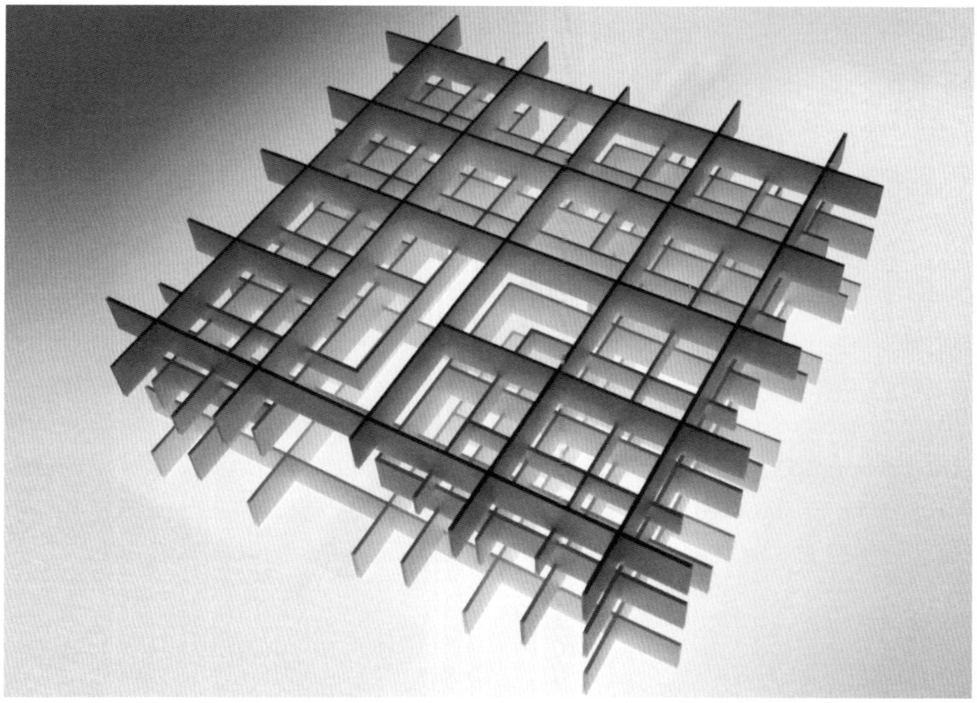

网格体系 Structural System

空间模型 Spatial Model

“无柱”空间
Open Space without Column

错位的墙面
Staggered Wall

叠合的书架墙
Overlapped Bookshelf Wall

开放的阅览空间
Open Reading Space

内敛的阅览空间
Enclosed Reading Space

5 历史文献阅览区 Historical Literature 古籍珍
4 政府信息公开查阅服务中心 Governme
3 音乐图书馆 Music Library
2 中文报纸阅览区 Chine
1 视障读者服务区

通廊式大厅
Corridor Space

最具象征性的空间是首层的通廊式大厅。它贯通南北，犹如连接中心湖公园和平江道的通廊。在这个通廊中，布置了总接待台和座凳，它既是整个建筑的中庭，还兼具多种公共用途，方便读者使用。面对中庭，周边的公共活动功能，如展厅、自习室、餐厅等等，能够自然地渗透进来。自然光透过上部的天窗漫射进来，形成了明亮柔和的光环境。此处虽然是室内空间，却如同街道般通透开放，让各种各样的人流穿梭交错其中。

The corridor on the first floor is a symbolic space. While running through south to north, it feels like a tunnel that connects Center Lake and Pingjiang Road. With reception desk and stools, the corridor becoming an atrium of the whole building with multiple functions to serve the reader conveniently. Functions such as exhibition hall, individual study room and dining hall facing the corridor forms an integration space. Skylight comes through the window softly floods the room with soft light makes a penetrating and thorough feeling while crowds of people pass by.

阶梯状的阅览平台
Stair-like Reading Terraces

阶梯状的阅览平台
Stair-like Reading Terraces

总接待台
Information Desk

总接待台
Information Desk

外墙采用石材百叶。在方案征集阶段，设计采用双层呼吸幕墙将建筑物包裹起来。在方案深化完善阶段，经过与规划设计组和文化中心其他建筑师共同商议，决定博物馆、美术馆、图书馆的外墙分别采用不同肌理的石材。对于图书馆，为了做出明快通透的效果，提出了石材百叶的方案。宽90mm、厚150mm的竖向花岗岩条，以190mm的间距布置。从斜向看，与一般石材外墙接近，从正面看则极具通透感。柔和的自然光透过石材百叶照射进阅览室内，创造了舒适的读书空间。到了夜间，室内的光线通过百叶散射出来，给人留下了深刻的印象。

Striped stone is being used at the exterior wall. At the scheme collecting phase, building is wrapped by ventilating glass panel wall. Rounds of discussion and negotiation between planners and architects have been made before deciding using various textures on Museum, Art Museum and Library during the modification phase. Striped stone have been adopted in the Library to keep the skylight penetrating effect. Spaces between each strip is 190mm with 90width and 150mm height marble bar. The purpose is to make the skylight penetrating effect in the front and a strong feeling for stone from the oblique angle. Comfortable reading spaces has been created with day lighting coming through the louver softly in daytime while room light comes out at the night to make a strong impression.

西侧 West View

西南角 Southwest View

沿湖立面　Lake View

沿湖夜景立面　Night View from Lake

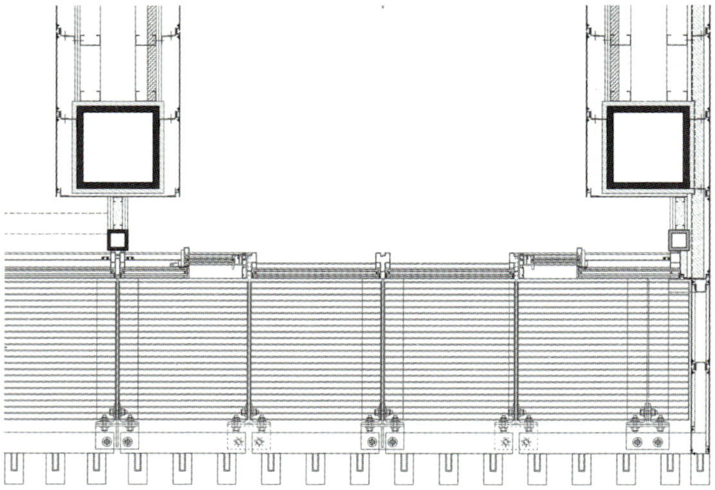

平面幕墙节点
Detailed plan of Curtain Wall

石材百叶幕墙
Stripe Stone Curtain Wall

局部立面
Partial Facade

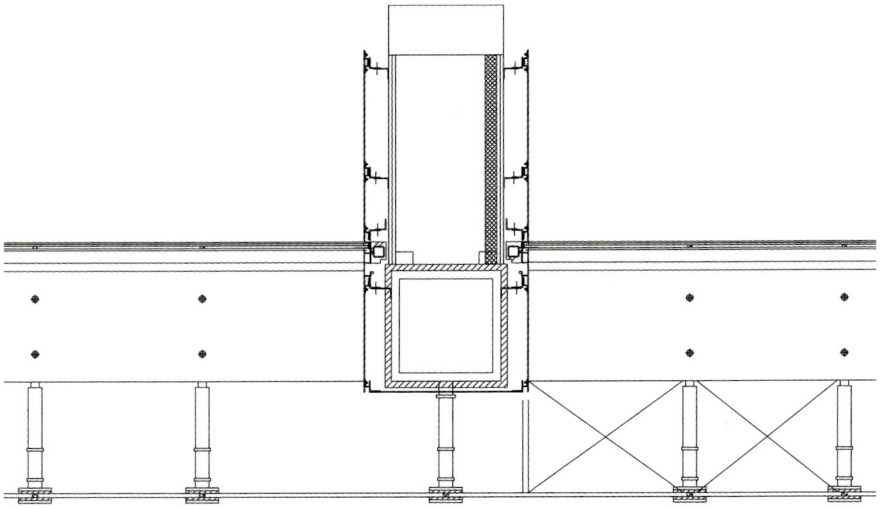

平面幕墙节点
Detailed plan of Curtain Wall

玻璃幕墙
Glass Panel Wall

大厅北侧入口
North Entrance

草坡裙房下的自习室
Individual Study Room Underneath Grass Slope

草坡裙房
Grass Slope Podium

内装设计极其简练。墙面重叠组合的特征作为建筑的要领被充分表现出来。除阅览室的墙面布置了书架外，其他墙面都是用白色的铝合金板作为饰面。墙面书架和铝合金板的分格对齐，均采用宽600mm、高340mm的规格划分，强化了整个墙面的统一感。铝板通过四周倒边，形成3种不同厚度的铝板单元，通过无规则地穿插布置，完成了整个墙面的丰富肌理，使得简洁的大幅墙面看上去并不单调。铝板的面层涂料略有光泽，具有一定的镜面反射效果。虽然室内的主色调是纯净的白色，却能够在视角、距离、光源变化之下，呈现出各式各样的微妙表情。与墙面相协调，顶面也采用白色的百叶吊顶。地面则采用深灰色，既营造了空间整体的沉静感，又能与各种家具及陈设相协调。在室内，以纯净的白色、灰色背景，色彩丰富的书籍与穿梭其间的读者被映衬出来，成为空间中动静相宜的两位主角。

The especially concise interior design stands out the complicated arrangement of walls. The entire wall uses aluminum alloy as decoration except for the reading room decorated with bookshelves. Bookshelves are aligned with the mesh division of aluminum alloy panel which is 600mm in width and 340mm in height to form an integrated feel. By randomly assembling different chamfer angle of aluminum panels, concise larger acreage wall is full of texture changes. Coating of aluminum panel has luster character and spectacular reflection. Although the interior theme is white, it shows various subtle expressions in all sorts of angle, light and distance. To cooperate with wall, the ceiling has adopted the same white stripe texture. With deep grey ground, assortment of furniture and harmonious displays, the whole space shows a calming and integration expression. Colorful books and reader linger in between books pops out since the interior are all in pure white and grey, which makes a congeniality atmosphere of reader and books.

空中书架
Floating Bookshelf

欢迎广大读者光临天津图书馆
欢迎广大读者光临天津图书馆

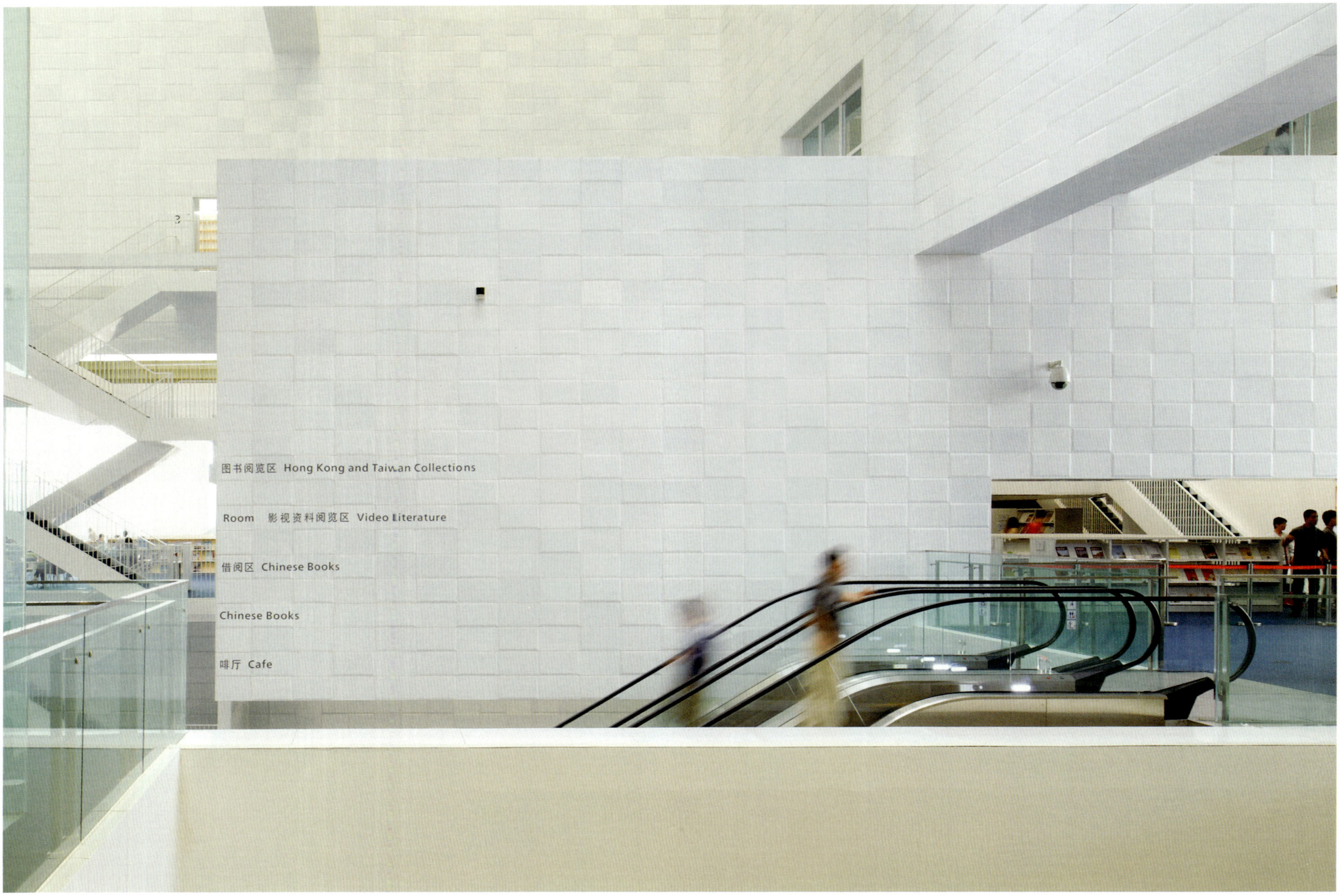

铝板墙　Aluminum Wall

书架墙 Bookshelf Wall

白色墙面
White Wall

白色吊顶
White Hang Ceiling

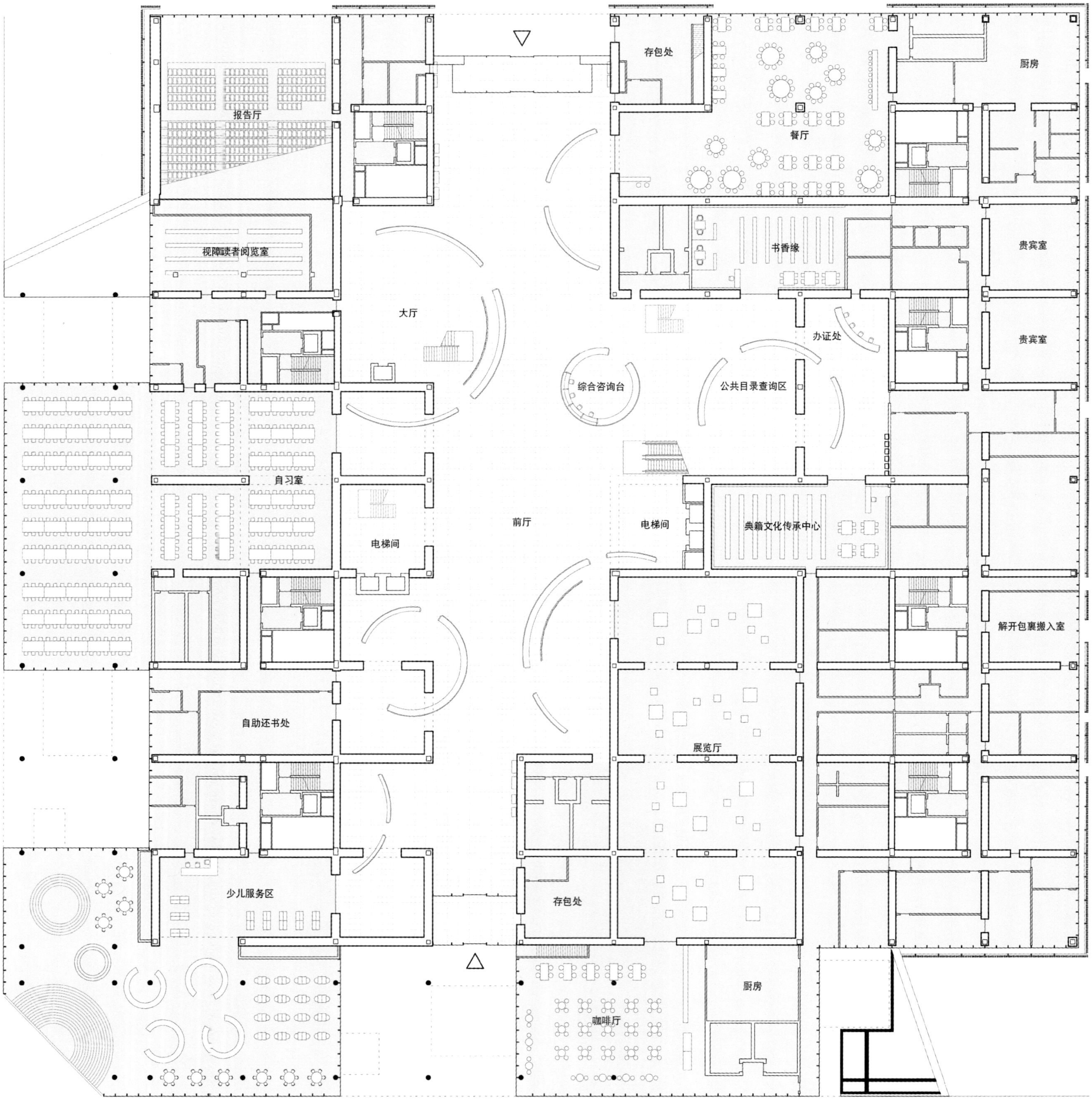

首层平面
First Level Plan

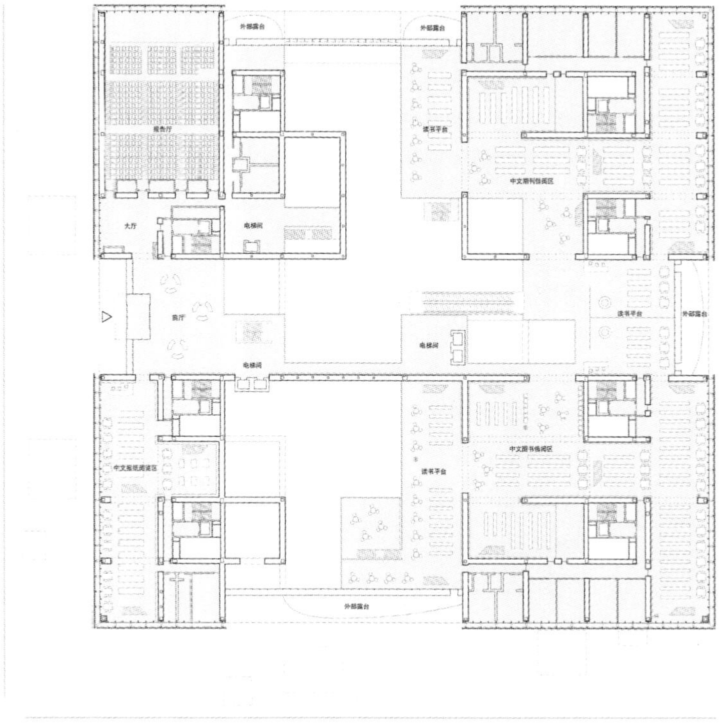

二层平面
Second Level Plan

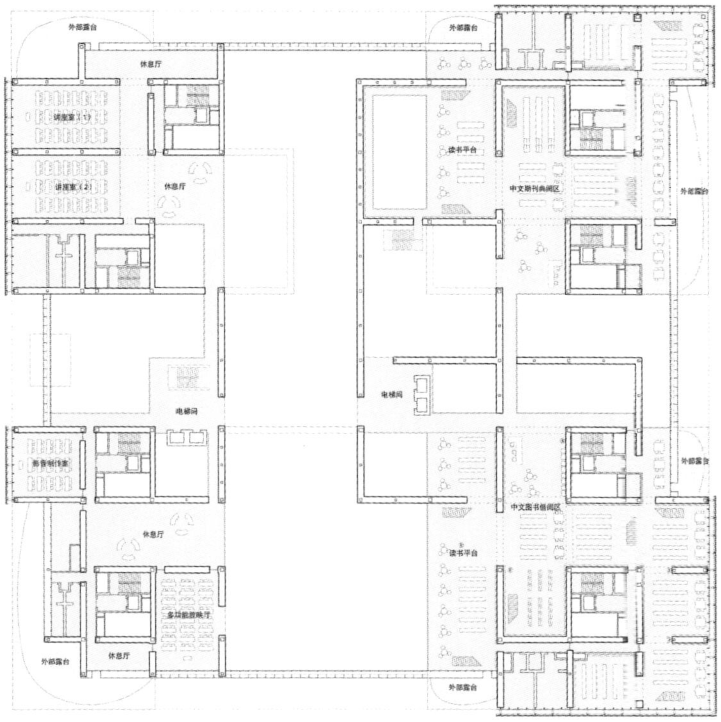

三层平面
Third Level Plan

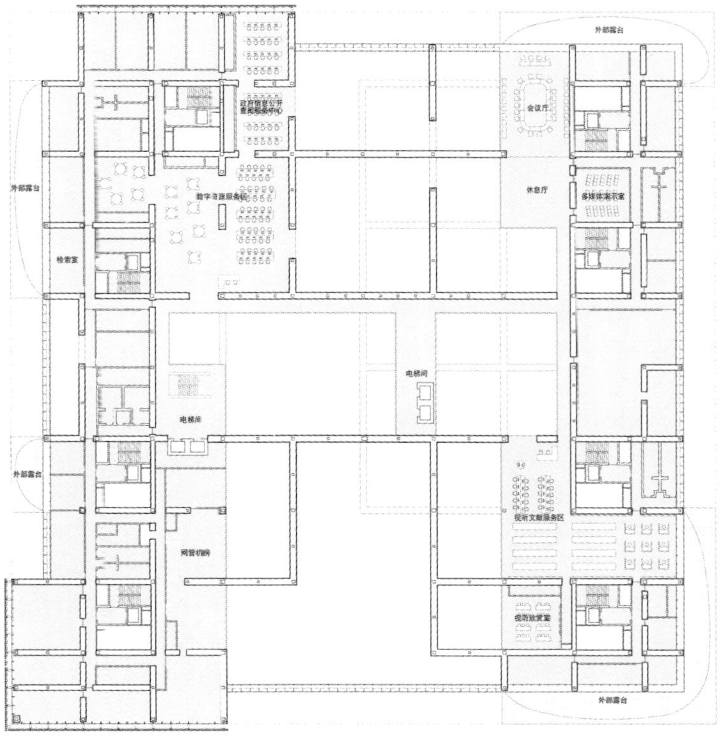

四层平面
Forth Level Plan

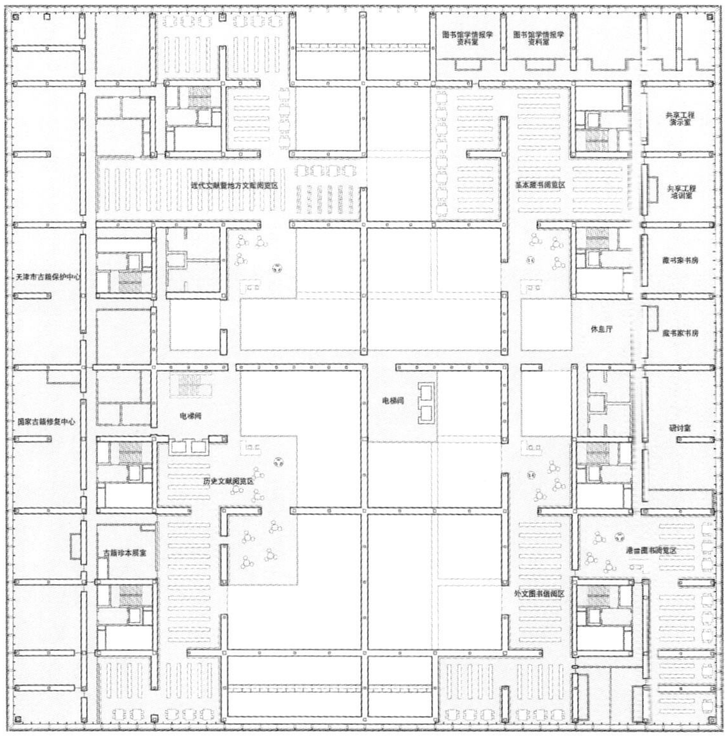

五层平面
Fifth Level Plan

结构设计

复杂建筑空间决定了本工程需要采用特殊的结构体系与节点形式：共享空间贯通南北，相邻层墙面错位布置；在121个柱位中，仅45根框架柱可以竖向贯通；各层楼板开洞面积均大于30%，楼层水平方向连接薄弱，一至三层形成了较为独立的四个部分；对梁、柱及桁架杆件普遍限制尺寸为500mm；室内外存在大量悬挑达10.2米的体块。

针对上述特点，采用了钢框架—支撑与空间桁架相结合的结构体系。利用8个交通核集中布置上下贯通的支撑，保证足够的抗侧力构件；在四、五层，以交错、叠放、吊挂、悬挑的方式布置了大量全层高桁架，形成空间结构受力体系，既承担竖向荷载也传递水平力，并加强了上部整体刚度。

本工程结构杆件尺寸小、板件厚、箱型截面多、杆件交汇多。为此，梁（桁架）柱节点采用了宽边带悬臂段隔板贯通等形式，空间桁架的多杆件交叉部位采用了矩形管铸钢节点，并进行了足尺及缩尺试验。对于大跨度及长悬臂悬挑区域，进行了多部位楼板的舒适度分析与控制。针对多项结构超限，在采用Midas Gen、ETABS、ANSYS等多个软件计算分析的基础上，进行了1：20结构模型的地震模拟振动台试验，试验结果证明设计达到预定的抗震性能目标。通过施工全过程模拟计算分析，设置了近百处临时支撑，结构设计全面满足施工荷载要求。

Structure Design

Complex architectural space determines special structural system for the project: shared space run through south to north; various wall arrangement between upper and lower level; only 45 of 121 frame columns could run-through vertically; punch area of floor slab is greater than 30%; horizontal connection between each floor is weak; The space between second floor to third floor form four independent partitions; The radius of beams, columns and truss rod are generally limited to 500mm; cantilever truss structure of 10.2 meters exist indoor and outdoor.

According to characteristics motioned above, steel frame and space truss structure system are combined to support the whole structure. Eight transport hub centralized arrangement up and down through the support ensure sufficient lateral force resisting members. Both bear the vertical load transfer horizontal forces and strengthening the upper part of the overall stiffness.

The engineering structures rod sizes are small; the thickness of the plates and box section are more than rods intersection. To this end, the beam (truss) column node broadside with cantilever section divisions through other forms. The space truss rod cross-site using the rectangular tube steel joint, and the full scale and reduced scale test. For large span and long cantilever cantilevered area, carried out the analysis and control of multi-site floor comfort. For a number of structural overrun, calculated using Midas Gen, ETABS, ANSYS software analysis based on a the 1:20 structure model shaking table test, The test results prove the design to achieve the seismic performance objectives. Through the whole construction process simulation analysis, we set up nearly 100 temporary supports to meet the full requirements of the construction loads.

施工现场 Construction Site

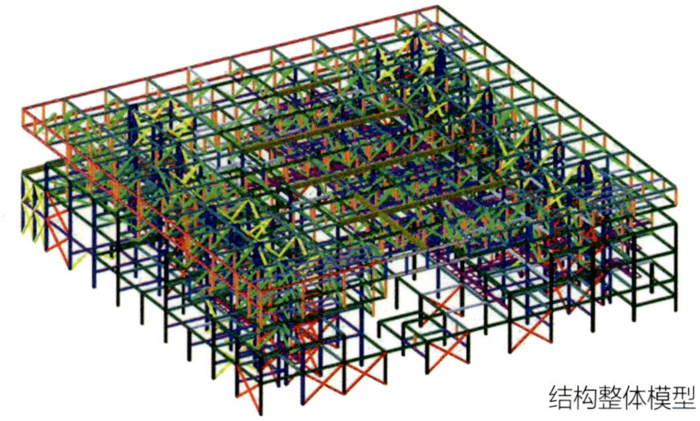

结构整体模型
Overall Structural Model

太悬挑结构
Cantilevered Truss Structure

暖通空调设计

结合建筑内部空间高大、贯通的特点，立足于室内环境热舒适与节能，图书馆公共空间主要采用地板送风空调系统，本系统由空气处理机组、地板静压箱、旋流地板送风口、回风排风系统组成。通过计算、性能分析，决定采用Ø200mm旋流地板散流器送风口，每个送风口的风量控制在150m³/h左右，送风温度不低于19℃，以免风口附近人员感觉到不舒适的冷风。在地板送风系统设计中，合理确定静压箱高度及其作用范围，是实现均匀送风的关键。通过CFD的模拟，确定静压箱净高为0.18m，保证均匀送风的送风量可达7000m³/h。同时，考虑到热力衰减因素，确定了从静压箱入口到房间散流器之间的合理输送距离为10~15m。

为提高空间利用率，充分利用辅助房间上部的夹层空间作为空调机房，并采用紧凑型空调机。出于节能考虑，在过渡季最大限度地利用室外新风，实现免费供冷，所有的全空气空调系统均设置了与空调送风量相对应的过渡季排风系统，以及与空调季新风量相对应的空调季排风系统。

HAVC Design

Combined with the large, thread characteristics of the inner space, based on the thermal comfort and energy saving, the under-floor air distribution system(UFAD) is adopted for the public space. The system is composed of air handling units, under-floor plenum chamber, swirl vertex floor outlet, return air and exhaust air system. By calculating and performance analysis, the swirl floor outlet of Ø200mm is adopted, the quantity of each air supply outlet is about 150m³/h, the temperature of supply air is higher than 19℃ to avoid making people near the outlet uncomfortable by the cold air. Determination of the height and range of plenum chamber is the key point in UFAD design. By CFD simulation, the plenum chamber height is fixed on 0.18m, so that the supply air quantity could be up to 7000m³/h when the uniformity of air supply is well. Meanwhile, because of the thermal decay, the distance from inlet of plenum chamber to outlet is ascertained to 10~15m.

To increase space utilization ratio, the mezzanine upon auxiliary room is fully used as air-conditioning room, where compact style air-condition is adopted. Considering of energy saving, the outdoor fresh air is used to achieve free cooling in transition season. The exhaust air part corresponds to supply air part in transition season and exhaust air part corresponds to fresh air part in air-condition season has been set up in all of the all-air condition system.

给排水设计

在空间高大处，设置智能型主动喷水灭火系统。在古籍书库、缩微胶片库、变电站等严禁水渍污染处，设置IG541气体灭火系统。其他常规空间设置喷淋系统。智能型主动喷水灭火系统与喷淋系统的消防水泵合用，合理地控制了造价。所有位置的自动喷淋系统均采用预作用式喷淋系统，避免因误喷或管道泄漏造成对书籍的损坏。共享空间、带夹层的阅览室、餐厅操作间的喷头采用快速响应喷头。

由于屋面美观的要求，屋顶不具备设置消防水箱的条件。设计采用气体顶压的方式取而代之，消火栓系统气压罐的容积为18立方米，喷淋系统气压罐的容积为10立方米。

Water Supply and Drainage Design

Intelligent and initiative sprinkler system are set up in the high room. IG54 gas fire extinguishing systems is set up in ancient stack room, microfilm storeroom, transformer substation and other places which strictly prohibit soaked by the water. Sprinkler system is set up in the other places. It can control the cost reasonably by using one suit of fire forcing pump in both intelligent and initiative sprinkler system and IG54 gas fire extinguishing system. Preaction sprinkler system is used in any place where setting sprinkler system to avoid the damage in case of a leak and false sprinkler. ESFR sprinkler head is used in the shareware room, reading room with interlayer and operating dining-room.

Because of the request to the roof handsome, there are no conditions to set the fire water tank in the roof instead of using the Gas pressure fire water device. The air pressure tank which can provide 18m³ water for fire hydrant system, and the other air pressure tank which can provide 10m³ water for fire sprinkler system.

智能化设计

天津图书馆是一座全数字化的智能图书馆：一个拥有多种媒体内容的数字化信息资源中心，能够为用户提供便捷的信息化服务；一个电子化信息的仓储中心，能够存储大量各种形式的信息，用户通过网络可以方便地访问。

计算机网络系统由内部应用网、公共外网、安防网组成。内部应用网为万兆骨干网。公共外网为图书馆读者及工作人员提供有线、无线Internet接入，为千兆骨干网。安防网为智能化系统各终端设备提供网络通讯平台，为千兆骨干网。

Intelligent Design

Tianjin Library is an all-digital intelligent library. It's an information resource center, with different types of digital media contents and can provide convenient services of a rich array of information and resources; it's an electronic information storage center, with different subject and large amount of information. User could easily access it through the internet.

The computer network system is composed of the internal network, public external network and security network. Core equipment room of internal application network provides ten thousands Megabits backbone network. Public external network provides wired and wireless Internet for library users and staff personnel which is a thousand Megabits backbone network. Security network provides network communications platform for intelligent system terminal device, which is thousand Megabits backbone network.

二层地板送风 Under-Floor Air Distribution System at Second Floor

阳光乐园 Sunshine Paradise

天津华汇工程建筑设计有限公司
Huahui Architectural Design & Engineering Co., Ltd.

建筑设计

（1）工程概况

本案位于文化中心东北隅，主入口轴线以对角线的方向指向中心湖面。结合来自乐园道和原青少年活 动中心的人流，设计为一个三叉形的内部公共空间，以此组织来自三个方向的使用者。同时，三叉形将建筑物内部分割为三个功能组团，可以各自独立运行。组团之 间又以三叉形的公共空间相连接形成统一的整体，共同组成了三叶草形的外部造型：其中，功能组团为叶片，公共空间为叶脉。

外墙立面形式结合三叶草形体，最终确定以石材为主体材料，配以金属板和玻璃，形成一种连续的、均质的图案式建筑表皮，以此保持形体的完整性和连贯性。其中石材被分割成宽度以一定模数逐渐变窄的5种条状板材，侧面镶以色彩变化的金属板。以5种标准板材为单元，构件按层数和相同的水平距离（1400mm），从上至下逐层由宽变窄进行排列，形成渐变的垂直缝隙的变化。再结合内部功能，设置不同高度的水平带窗，在窗前覆盖厚150mm、间距1500mm的垂直格栅，在窗前创造了一个连续的水平通透介面，它随着窗的高度变化，犹如一条上下跳跃的声音波形，缠绕在三叶草形的连续曲面之上，使其与青少年的性格相一致，充满生气和活力。

（2）中庭设计

位于三叶草中间的三叉形共享中庭，为整个建筑的最主要交通组织空间。为保证中庭空间的完整、流畅和灵活，将防火卷帘沿1-4层中庭回廊外侧设置，将中庭及回廊作为一个防火分区进行设计。对于由此产生的防火分区面积超大及中庭首层疏散距离超长的问题，则通过消防性能化设计进行解决，并采取自动扫描射水高空水炮灭火系统（智能水炮）、智能疏散系统和高大空间的火灾监测系统等消防加强措施。

（3）屋顶采光顶系统

在 中庭屋顶设置自然光导光管系统，将自然光收集，并传送到中庭内部。自然光导光管系统主要由集光器、导光筒、漫射器三部分组成。集光器让尽可能多的自然光进 入，同时防止灰尘、雨和紫外线的侵入，导光筒将自然光传输到需要的场所，漫射器将光线均匀地分布到室内，为空间提供照明。

摄影：魏刚
Photographer: Wei Gang

主入口
Main Entrance

阳光乐园全景 Complete View of Sunshine Paradise

外墙细节
Exterior Details

东南侧全景
Southeast View

外墙细节
Exterior Details

入口细节
Entrance Details

Architectural Design

（1）General Information

Located in the southeast corner of the cultural center, the axis of the main entrance in the design diagonally points to the central lake. Considering the traffic from the Leyuan Street and the original Youth Activity Center, an internal public space in the trigeminal shape is designed to organize users from three directions. The internal building space is divided into three sections which can operate independently. The three sections connect to each other and form an integrated entity, displaying an external clove-shape structure. The function sections serve as leafs and the public space serves as vein.

Considering the clove shape, the exterior wall façade uses stone as main materials coupled with metal plates and glasses so that a continuous, even pattern appears on the face of the building to keep the form integrity and continuity. Stones are classified into five strip types the width of which decreases based on a certain modulus, with colorful metal plates set on the lateral side. These standard plate structure materials are laid out based on the number of layers and the same horizontal distance (1400mm) with width narrowing from bottom to top, forming a gradual vertical gap with changes. Combined with the internal

室内中庭全景 Interior Atrium View

functions, horizontal ribbon windows in varied height are opened and vertical grids 150mm thick are placed in front of the horizontal ribbon window at intervals of 1400mm, creating a continuous horizontal transparent interface. As the window height changes, a fluctuating sound ware pattern unfolds in one's mind, winding on the continuous curve of the clove, which empathize with the youth personality and is imbued with vitality and vigor.

(2) Atrium Design

Located in the middle of the clove, the shared herringbone atrium is the main traffic and organization space for the whole building. To ensure the integrity, fluency and flexibility of the atrium space, fire resisting shutters are set along the first through fourth atrium and outside their corridors, and are designed as fire compartment. Consequently, the fire compartment area is large and the evacuation distance in the atrium on the first floor is long. To resolve this problem, performance-based fire design is adopted with fire controls measures, such as the high pressure squirt water gun with automatic scanning set within the atrium (smart water gun), evacuation system and fire monitoring system for high and large space.

(3) Roof Natural Light System

The roof in atrium shall adopt the natural light-pipe system which can collect natural light being transmitted to the inner space of atrium. The natural light-pipe system is composed of optical collector, light tube and diffuser. The optical collector can make more natural light enter as possible, at the same time, prevent the dust, rain and ultraviolet from intruding. The light tube can send the natural light to the required space. The diffuser can be used to distribute the natural light evenly in the house and provide lighting in space.

室内细节
Interior Detail View

室内中庭
Interior Atrium

中庭天光细节
Skylight Details

平面图
Plan

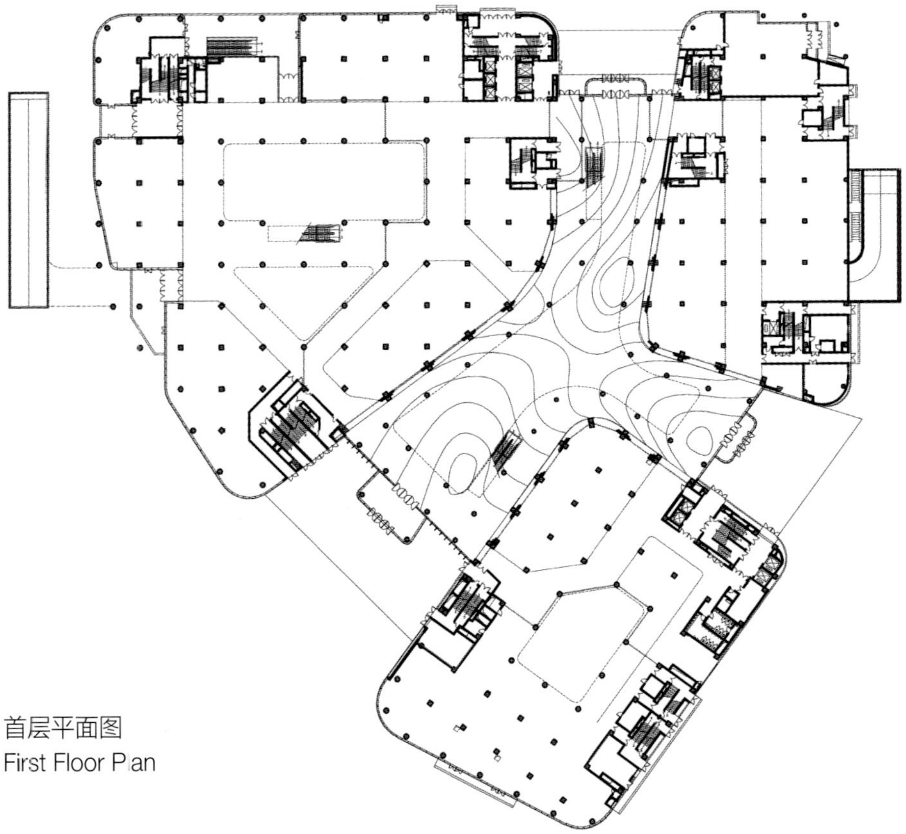

首层平面图
First Floor Plan

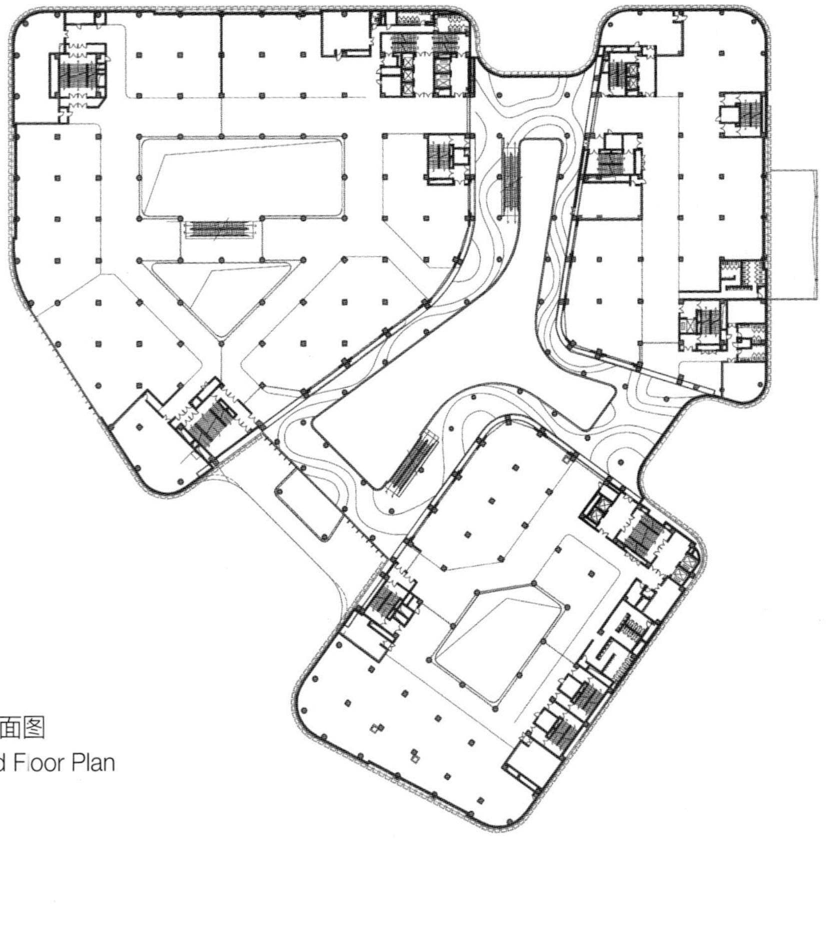

二层平面图
Second Floor Plan

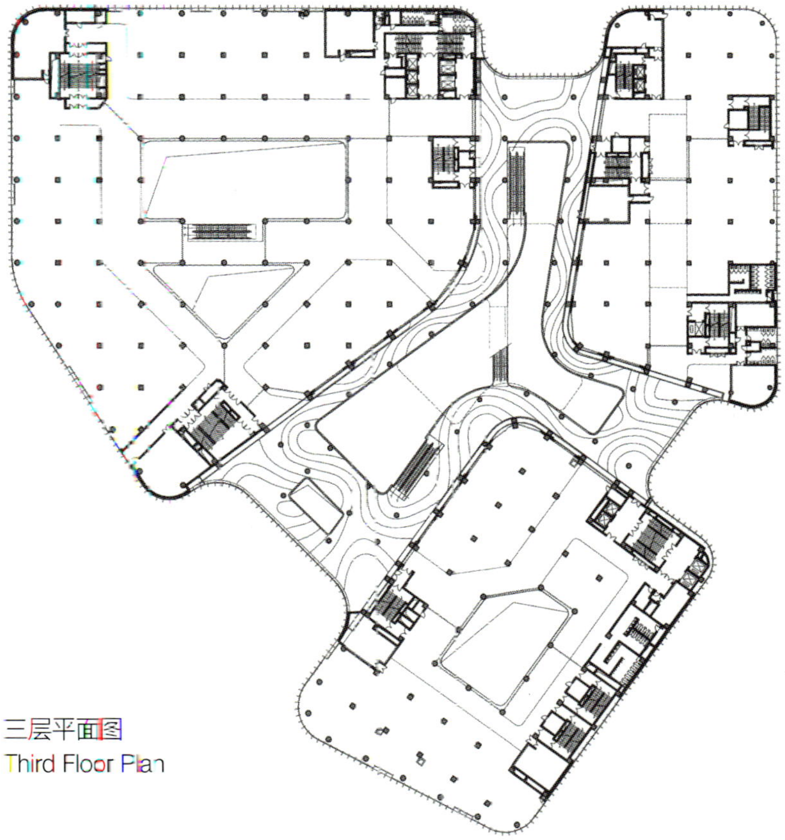

三层平面图
Third Floor Plan

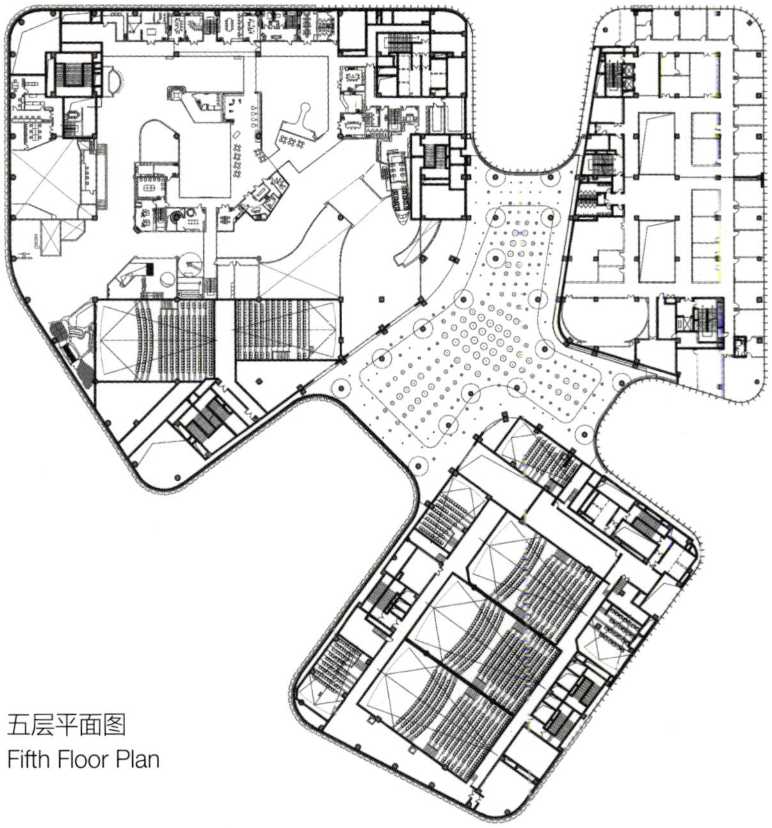

五层平面图
Fifth Floor Plan

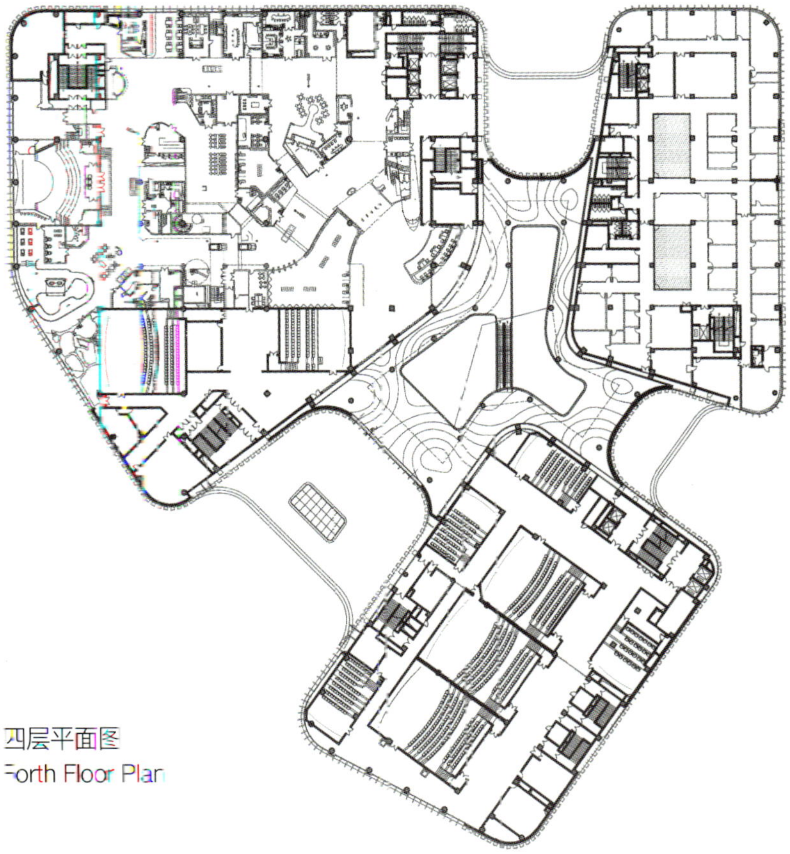

四层平面图
Forth Floor Plan

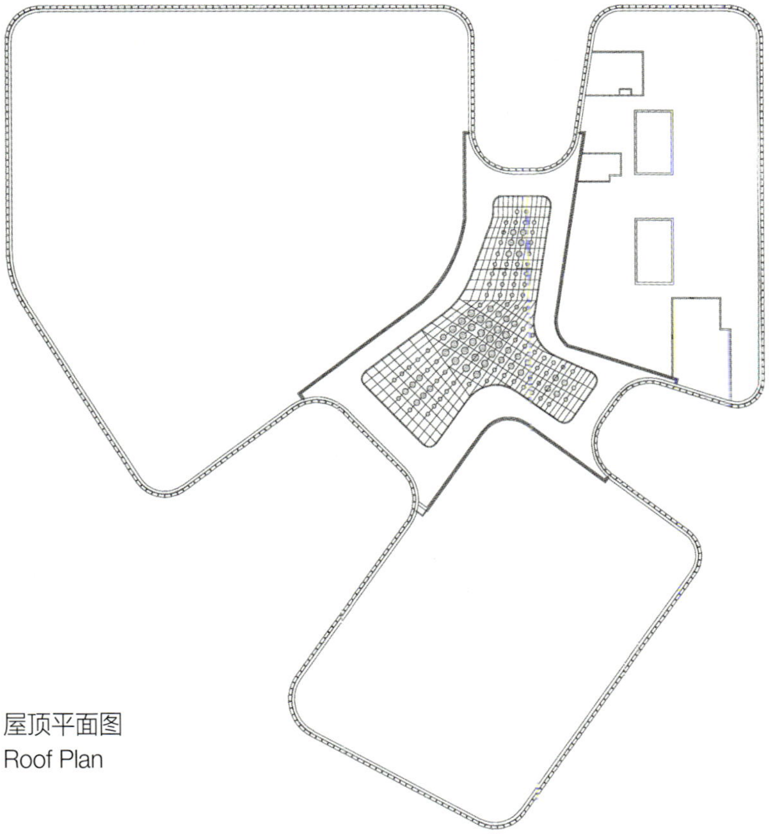

屋顶平面图
Roof Plan

立面图 Elevations

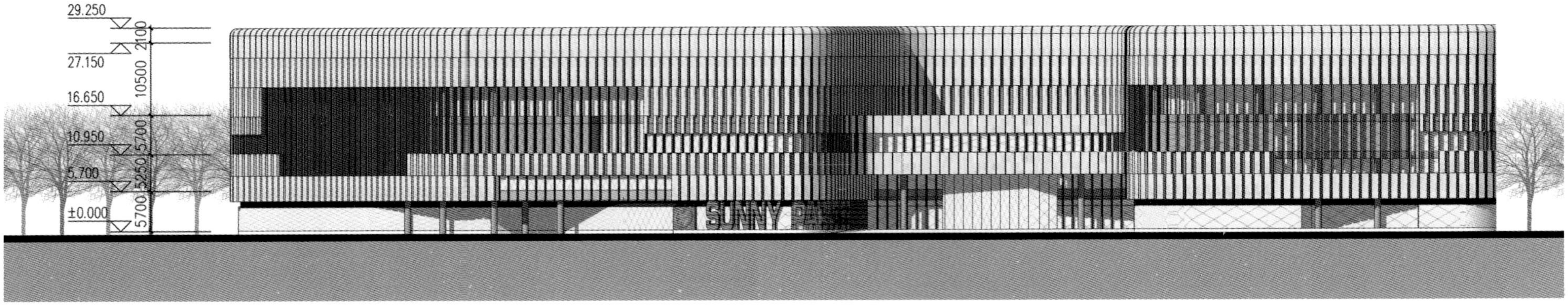

立面图一 Elvation-01

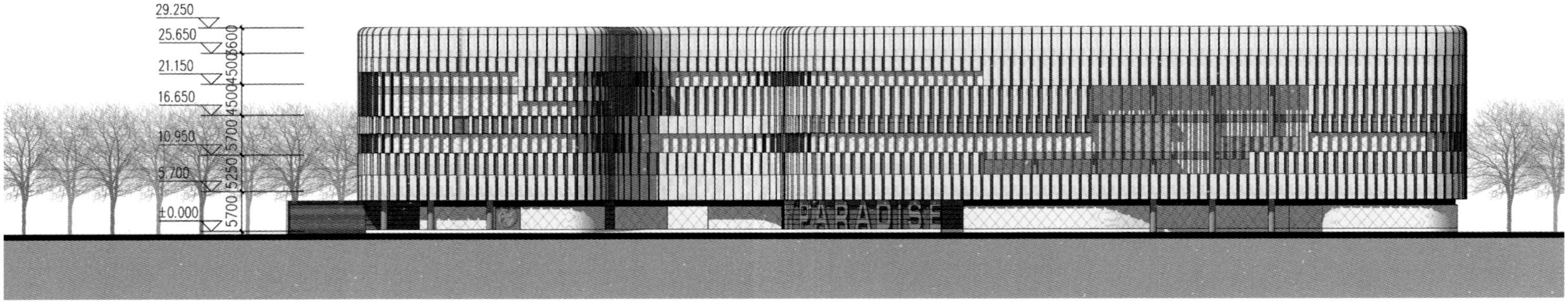

立面图二 Elevation-02

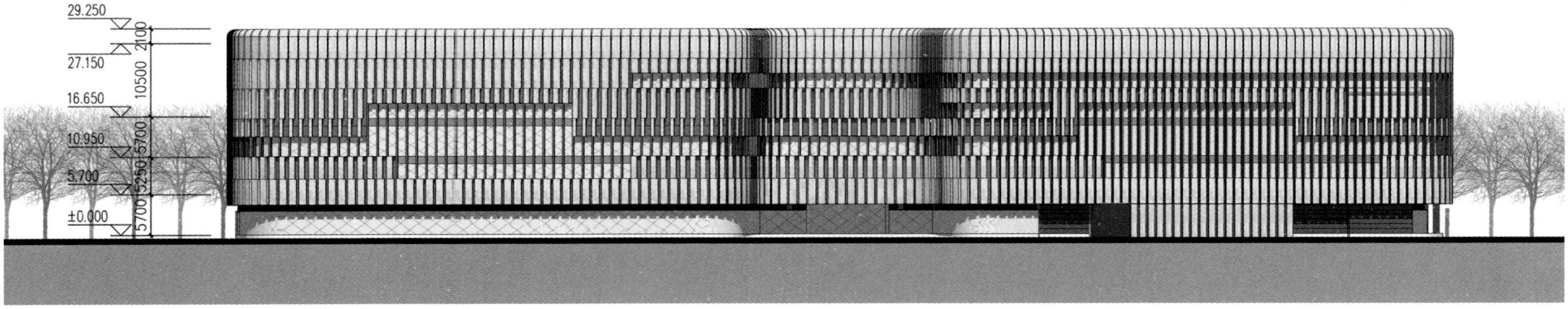

立面图三 Elevation-03

剖面图 Sections

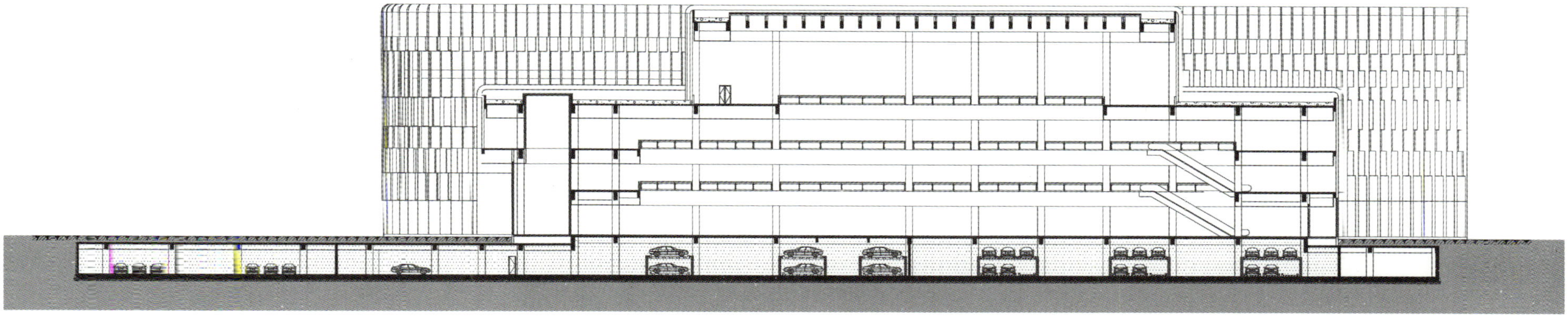

剖面图一 Section-01

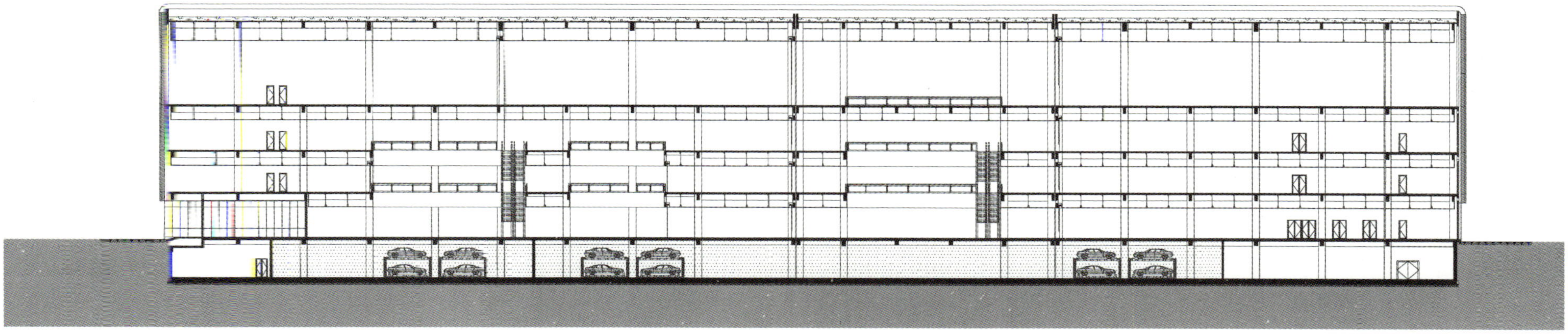

剖面图二 Section-02

建筑构造 幕墙系统

本项目幕墙系统复杂多样，包括框架玻璃幕墙系统、石材幕墙系统、首层框架幕墙系统、主入口拉索幕墙系统、采光顶系统、室内玻璃肋系统等。

石材幕墙系统

建筑外立面大面石材突出玻璃面，且石材面板宽度呈现有规律的变化，从顶部往下和从底部往上均变窄。为实现此建筑效果，建筑采用热浸镀锌钢架支撑，石材面板由不锈钢背栓固定于钢架上。钢架要求在工厂加工焊接完成后在现场整体安装。要求幕墙装饰条石材两侧均转100mm厚度，以从视觉上增加石材厚度。

首层框架幕墙系统

首层幕墙采用钢框架支撑的玻璃幕墙系统，所有外露面均包覆不锈钢板，玻璃面板采用中空超白低辐射玻璃，以实现建筑首层通透性。

主入口幕墙提供

主入口为三层通高带采光顶的空间。为实现建筑入口通透光亮效果，立面采用竖向拉索玻璃幕墙系统，玻璃面板通过不锈钢夹具固定到张拉索具上；屋顶采用玻璃面板和张拉弦钢结构支撑体系；玻璃均采用中空超白低辐射玻璃。

Architectural Construction Curtain Wall System

Stone Curtain Wall System

The stone outstanding glass surface of building exterior façade is wide at two ends and narrow in the middle part with regular variation in its width of stone panel. To achieve this effect, this building shall apply the hot-dip galvanized steel frame as support while stone panel is fixed to steel frame by stainless steel back bolts. The steel frame is required to be installed onsite after completing the plant processing and welding. It is required that the thickness at two sides of stone trim strip of curtain wall is 100mm and adds the thickness of stone panel from vision.

First Level Frame Curtain Wall System

The first-storey curtain wall shall adopt glass curtain wall system which is supported by steel framework. The entire exposed surface covers stainless steel plate and glass panel shall employ the glass which is featured by the hollow center and super white and low radiation to realize the permeability of first-storey of the building.

Curtain Wall of Main Entrance

The main entrance refers to the space for sunroof of 3 storeys. To achieve permeability and light effect at entrance of building, the façade shall apply the vertical-cable glass curtain wall system and the glass panel is fixed to tension rig by stainless steel fixture; the roof shall adopt the glass panel and tension string steel-structure supporting system; the glass shall apply the glass which is featured by the hollow center and super white and low radiation.

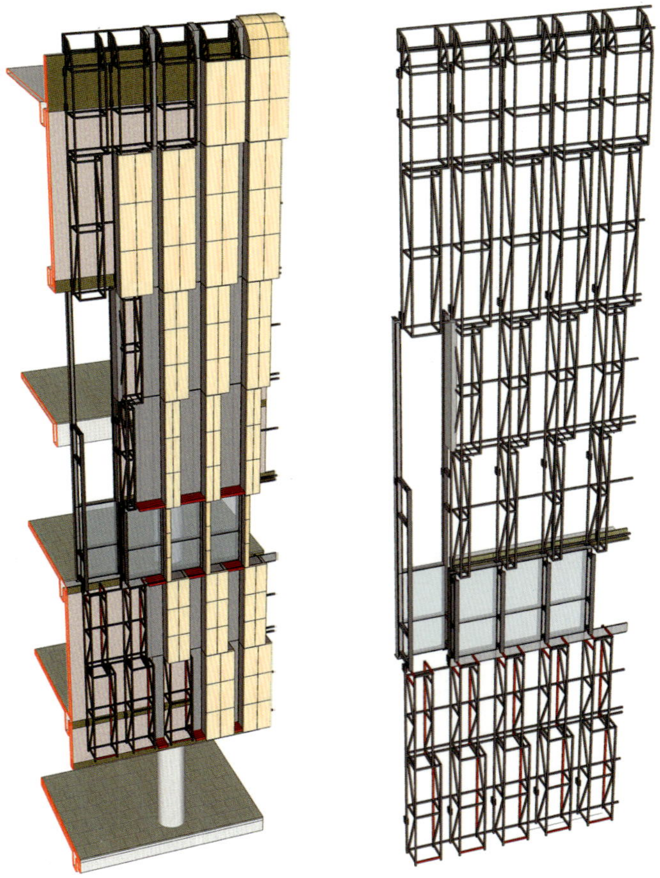

幕墙构造示意图
Diagram of Curtain Wall

幕墙构造模型 Detail Physical Model of Curtain Wall

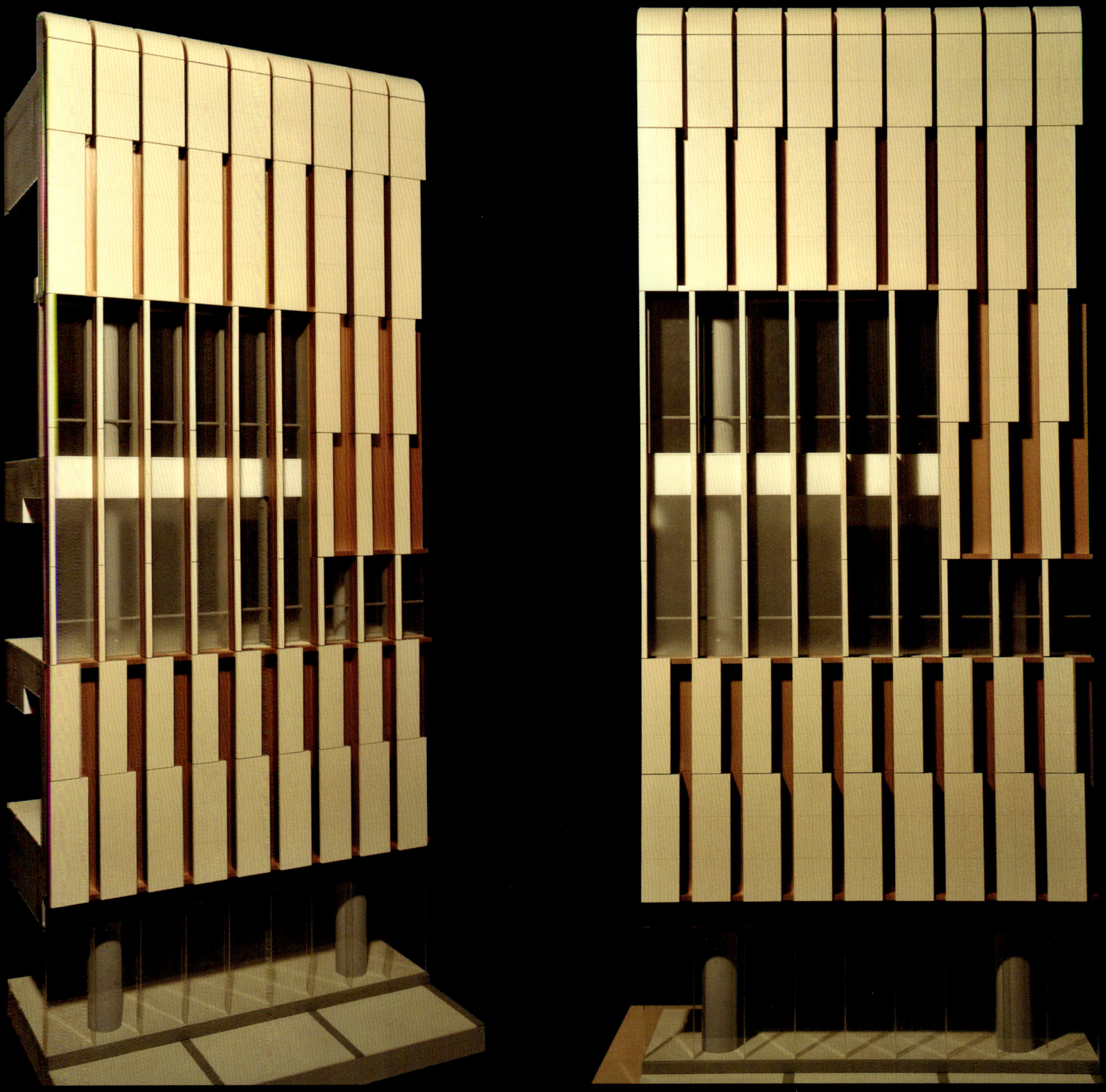

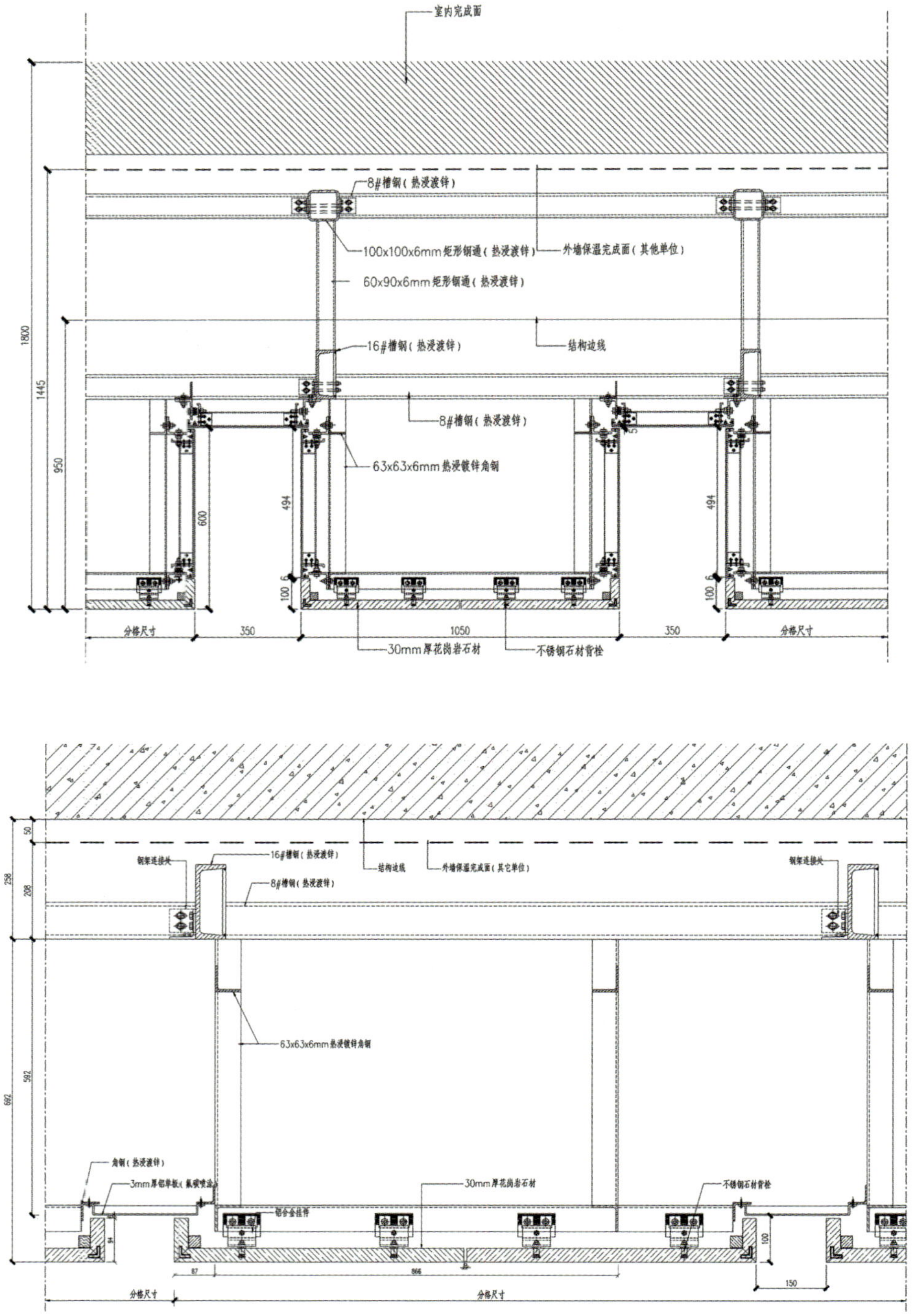

幕墙构造详图 Construction Drawings of Curtain Wall

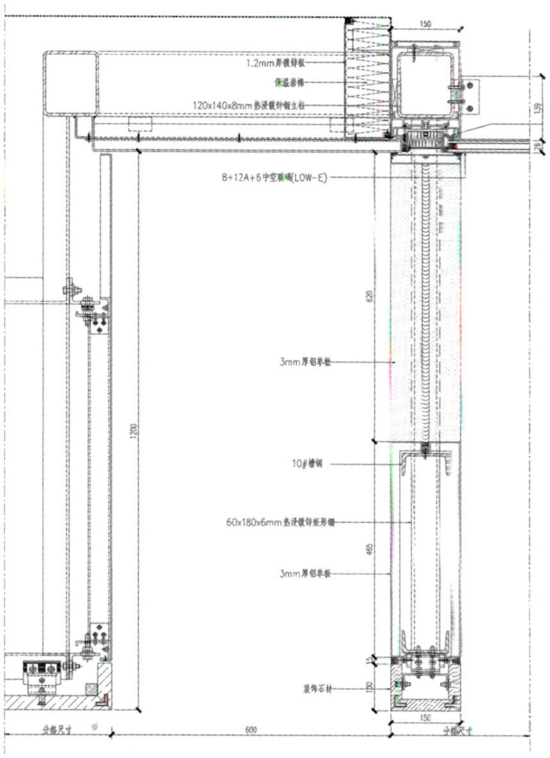
150
1.2mm厚镀锌板
保温岩棉
120x140x8mm热浸镀锌钢立柱
139
26
8+12A+6中空玻璃(LOW-E)
620
3mm厚铝单板
1200
10#槽钢
60x180x6mm热浸镀锌矩形钢
465
3mm厚铝单板
15
100
装饰石材
150
分格尺寸
600
分格尺寸

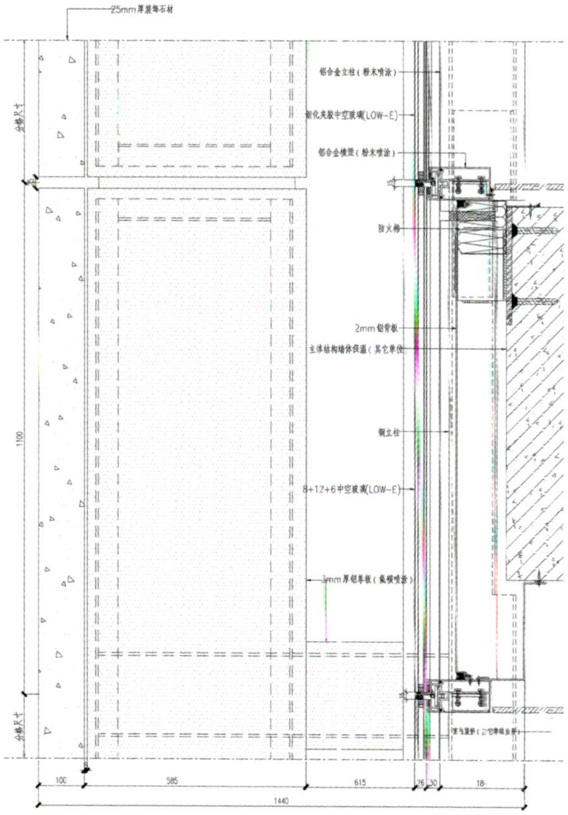
25mm厚装饰石材
铝合金立柱(粉末喷涂)
钢化夹胶中空玻璃(LOW-E)
铝合金横梁(粉末喷涂)
防火棉
2mm铝背板
主体结构墙体保温(其它单位)
钢立柱
8+12+6中空玻璃(LOW-E)
3mm厚铝单板(氟碳喷涂)
分格尺寸
1100
分格尺寸
100
585
615
26
30
18
1440

结构设计

本工程地上四层，局部五层，地下一层，主体结构最大高度27.15m。本场地抗震设防烈度为7度，设计基本地震加速度值0.15g，设计地震分组为第一组。建筑抗震设防类别为乙类，设计基准期为50年。根据使用功能及结构特点，本工程地下，地上结构均采用钢筋混凝土框架结构。

根据该工程方案设计概念，上部主体分为大叶，中叶，小叶三个功能区，中间通过走廊连为一个整体。大叶，中叶和小叶三个功能区位于中间走廊的三个方向，形成三角行布置,且三个单体不同角度不规则布置，中间走廊连接较为薄弱，地震时容易开裂，造成不安全因素。在平时温度应力作用下，也易形成温度裂缝,如将上部单体全部连为一个整体,则需要在楼板，梁内布置预应力钢筋，加强薄弱部位的构造，还需对此进行专门超限结构分析，此做法对于整个工程的造价和施工工期均有较大影响。因此在满足使用要求的基础上,将上部单体按照上部功能的要求，分为大叶，中叶，小叶和中间走廊四个结构单体，单体间设置抗震缝,四个单体分别独立分析和计算，有效减少工程造价，减少施工工序，缩短施工工期，以简单的方法同时满足了安全，经济和使用方便的要求。

Structural Design

According to the design concept in this project, the upper body can be divided into 3 functional domains, namely big lobe, middle lobe and small lobe, forming to an entity by corridor in the middle. The big lobe, middle lobe and small lobe is situated in 3 directions in the middle of corridor, forming into triangle layout, moreover, 3 lobes in different views has irregular layout and the connectivity in the middle corridor is so weak that it can easily break when the earthquake occurs. Thereby it is an unsafe factor. As affected by stress under normal temperature, it is very easy to form a temperature fissure. Provided that all the lobes forms into a entity, it is required that the prestressed steel should be arranged to floor and beam to strengthen the structure of weak part and the analysis on special transfinite structure should be conducted. This practice has great influence on the whole project cost and construction period. Therefore, after meeting the use requirement, the upper monomer can be divided into big lobe, middle lobe, small lobe and middle corridor as required by the above functions. Each monomer is equipped with seismic joint and conduct the independent analysis and calculation respectively so as to reduce the project cost effectively and construction process and narrow down the construction period, meanwhile, satisfy the requirement for safety, economy and convenience to use in a simple manner.

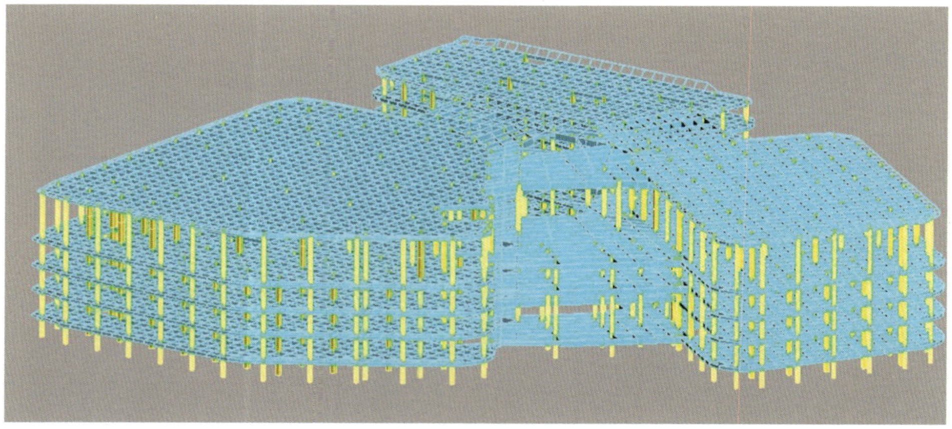

结构整体模型 Overall Structure Model

暖通空调设计

本建筑属于大空间建筑，具有内、外分区明显，内热负荷大等特点。对于靠近外围护结构的外区部分，直接受外围护结构日射得热、温差传热、辐射换热和空气渗透等因素影响，其空调负荷包括外围护结构冷负荷或热负荷以及内热冷负荷，天津气候条件决定外区有时需要供热有时需要供冷。内区不受外围护结构的日射得热、温差传热和空气渗透等影响，全年仅有内热冷负荷，其随区域内照明、设备和人员发热量的状况而变化，通常全年需要供冷。结合本工程特点，空调采用全空气定风量系统与风机盘管相结合的方案，由全空气定风量系统负担内区负荷及新风负荷，风机盘管负担外区负荷。可以很好地解决冬季内、外分区分别供冷、供热的问题，满足舒适度要求。同时由于全空气系统仅负担内区负荷，所以组合式空气处理机组的风量及设备尺寸会减小，相应的可以减小空调机房面积，增大有效使用空间面积；同时空调风管的截面积会减小，可以降低建筑层高，节省土建费用。

本建筑具有常年内热负荷大的特点，在过渡季采用全新风运行，利用室外空气降低室内温度，不仅可以减少冷机的开启时间，节省能源及运行费用，同时可以大大的改善室内的空气品质；冬季可以充分利用室外新风的冷量来消除室内余热，节能效果显著。

HVAC Design

This building is very spacious featuring stark difference between internal and external and large internal thermal load. As regards the external area near the outer-building envelope, it is fully exposed to the heat insolation, thermal transfer as result of different temperatures, radiation heat transfer, air filter, etc. Thus its air conditioning load includes cooling or heating load of outer-building envelope and internal cooling load. To be coupled with climate condition in Tianjin, it requires heat supply and cold supply alternatively. The internal area, on the contrary, is immune to the factors that affect the external area. It has only internal cooling load all year round and varies on the heating value radiated from lighting, equipment and people in this area. Therefore the internal area only requires cold supply. Based on this analysis, its air conditioning adopts all-air fixed air volume system combined with fan coil unit. The all-air fixed air volume system is responsible for the internal area load and cooling load from outdoor air, whereas the fan coil unit is in control of the external area load.

Due to its large internal thermal load, this building runs on all fresh air in transition season. Drawing on the outdoor air to cool the internal can reduce the running time of refrigerator, save energy as well as cost, and improve the internal air quality by leaps and bounds. In winter, outdoor fresh air can be maximally utilized to clear up the excess heat, saving energy remarkably.

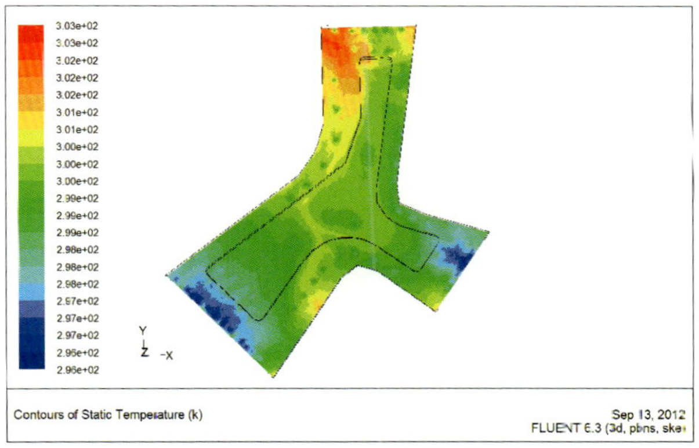

共享中庭 1.5m 温度分布云图
Thermal Analysis of Arium at 1.5m

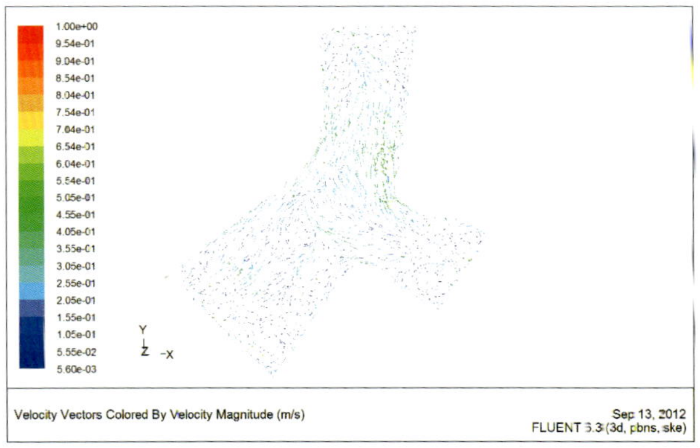

共享中庭 1.5m 高度横截面的速度矢量图
Vantilation Analysis of Arium at 1.5m

给排水设计

本建筑物为一个三叉形的内部公共空间，以此组织来自三个方向的使用者。同时三叉形将建筑物内部分割为三个组团，各自可以独立运行。组团之间又以三叉形的公共空间相连接形成统一的整体，共同组成了三叶草形的外部造型：功能组团为叶片，公共空间为共享叶脉，三叉形的共享空间首层至顶层通高，屋顶设置采光顶。

针对上述特点，建筑物共享中庭区域，高度超过普通自动喷水灭火系统的保护高度，自动喷水灭火系统采用自动扫描射水高空水炮灭火系统。该灭火系统，是针对现代大空间建筑的消防需要，采用新技术而研发生产的系统化消防产品。产品把红外传感、计算机、机械传动、远程通讯等技术集于一身，实现了高度智能化的现代消防理念。其主要特点是：全天候的自动火灾监控，全方位的主动喷水灭火。该系列产品还具有较强的电子电路和机械传动组件的自检能力，可迅速发现故障并报告消防控制中心。

建筑物按空间布局配置自动扫描射水高空水炮灭火装置。该系统与自动喷水系统合用消防水泵、消防水池、屋顶水箱，单独设置水流指示器、信号阀及模拟末端试水装置。尽可能降低工程造价。与其他方式自动喷水灭火系统相比，在满足消防要求的前提下，该系统设置喷头的数量少，最大限度展现中庭的整体效果。

Water Supply and Drainage Design

This building is a three-pronged internal space, providing room for users from three directions. Therefore this building is divided in three parts, each running on its own. These three parts are also linked, making a clover-shaped external styling: each part is a blade while the linkage is the vein. The space rises from the bottom to the transparent top without any hindrance.

Height of the atrium obviously exceeds the protection height covered by ordinary automatic sprinkler system. Taking all into consideration, the sprinkler system adopts automatic scanning water jet and water gun system. This system is a new-tech fire extinguishing system used exclusively in modern spacious buildings. Integrated with infrared sensing, computer, mechanical drive and telecommunications, this system is highly intelligent and sophisticated. It can monitor fire 24/7 and preemptively jets water. This system also boasts its self-checking capability in electronic circuit and mechanical drive. If anything goes wrong, it can immediately spot it and report to the fire control center.

This automatic scanning water jet and water gun system shares the fire pump, fire pool, roof tank, single set flow indicator, signal valve and simulated end water-test device with automatic sprinkler system, bringing the cost down as much as possible. With basic needs for fire fighting as prerequisite, number of nozzles are reduced to maintain the aesthetical effect.

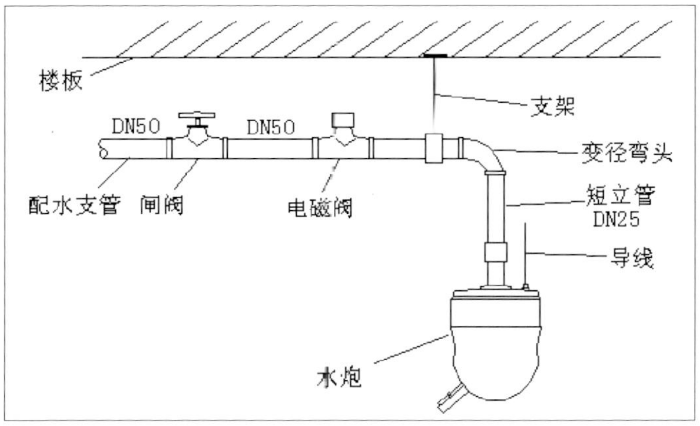

喷头安装示意图
Diagram Of Sprinkler Installation

电气设计

（1）智能疏散系统设计

阳光乐园作为青少年活动场合，对安全性要求很高。其内部中庭4层通高，作为一个防火分区，分区面积和疏散距离均超防火规范。因此，火灾时保证青少年快速撤离是本次设计的一大难点。本设计采用智能型疏散指示系统，该系统在消防控制室设置一台主机，在疏散通道和安全门设置智能型疏散指示灯，每个疏散指示灯都有一个唯一的地址编码，标志灯通过485总线与系统主机连接。火灾时由主机控制使标志灯双向箭头中的一个指向远离火灾点的箭头闪亮，指导人员进行远离火灾的安全疏散。

（2）高大空间的火灾监测

阳光乐园内部存在多处共享区域，共享高度超过感烟探测器的有效保护高度，为了有效探测共享区域火灾情况，本设计在采用红外光束感烟探测器保护共享区域的基础上，还采用了空气采样烟雾探测装置。该探测器安装在共享区域的吊顶内，采样管均布在共享区域吊顶内，由吸气泵通过采样管对防火分区内的空气进行采样。主机里面的激光枪分析空气的成分，得出空气中烟雾粒子的浓度。如果超过预定浓度，主机进行报警。

（3）电气火灾监控系统

阳光乐园内部集青少年活动场所、餐饮、办公、影院多功能于一身，业态多样，供电系统复杂，电源箱和控制箱达七百多个。而电气故障又是火灾发生的主要原因。有鉴于此，本设计采用电气火灾监控系统，探测装置安装在楼层总配电箱进线处，监控各楼层配电箱漏电流情况。如果超过设定值，主机进行报警。

Electrical Design

（1）Intelligent Escape System

Sunshine Recreation Center is the rendezvous of teenagers, a place where is required of high security. This center is a four-storey from-bottom-to-top atrium, exceeding the fire precaution requirement placed on the area and evacuation distance. Therefore quick evacuation in case of fire is a difficult task. This design applies intelligent evacuation indication system. A mainframe is set up in the fire control room and indicator lights are installed in escape route and at emergency exit. Each light is uniquely coded and linked to the mainframe via Bus 485. In case of fire, the mainframe sends signal to indicator light which, in turn, shows the direction for escape.

（2）Fire Monitor In Shared Areas

There are multiple open shared areas in Sunshine Recreation Center. The height is in far excess of the protection covered by smoke detectors. To better monitor fire, this design adopts air sampling smoke detector, in assistance with its infrared beam smoke detector. Air sampling smoke detector is installed in the ceiling of open shared area. Aspirator pump take the air sample through sampling tube. Laser gun in the mainframe analyses the elements in air and measures the concentration of smoke particles. If the concentration is higher that that is preset, the mainframe alarms automatically.

（3）Fire Monitor System

Sunshine Recreation Center is a multi-functioned area covering playground, dining place, office and cinema. Its complicated power supply system manifest itself in 700-odd power boxes and control boxes. Electrical failure is the primary cause of fire. Given this, this design applies electrical fire monitor system. The detector is placed at distribution box at each floor, monitoring leakage current. If the leakage current exceeds that is preset, the mainframe alarms automatically.

市民广场 Galaxy Mall

美国tvs design设计公司
tvs design
天津市建筑设计院
Tianjin Architecture Design Institute

建筑设计

天津市民广场商业综合体的设计理念源于对和谐生活的向往，意在创造出一片生活空间，一个可以让心灵、身体和精神同时得到满足的平衡的生活环境。一个融自然与文化、时尚与娱乐、社交与沉思于一体的和谐共存的场所。设计的主旨在于商业综合体应该呈一种开放的姿态去迎接文化中心的文化活动和自然景观。超越于一般的建筑，商业综合体和联系天津大礼堂与大剧院的花园步道交织融汇在一起。

Architectural Design

The design concept of Galaxy Mall originates from the desire of a harmonious life, and intends to create a living space, a balanced environment which can give people satisfaction in soul, body and mind, and a peaceful coexistence place where nature and culture, fashion and entertainment, social interaction and meditation expectation are integrated. Surpassing other common buildings, the commercial street and walking avenue linking the auditorium and the grand theatre are interwoven together. The major idea of this plan is that commercial complex should be an open attitude to meet the cultural activities and the natural landscape of the Cultural Center.

摄影：刘东、凌海
Photographer：Liu Dong, Ling Hai

南立面
South View

南立面实景图
South View

毗邻阳光乐园，在中心湖北侧，天津市民广场商业综合体将为天津带来高品质的商业体验，同时又为已有的文化中心规划锦上添花。该项目提供了约35万平方米商业建筑面积，功能包括零售、超大型自助商场、饮食、娱乐。建筑的尺度和周围的文化建筑相呼应协调，限高30M。一至五层层高为5.3M到6M之间，并设有地下一、二层，高分别为6.5M、4.5M。对于建筑体量、材料选择，重点考虑将大剧院、阳光乐园和商业综合体一起形成优雅和谐的整体。商业综合体设计的构想是为游客和天津的居民提供一个完整完全的商业娱乐体验，而不仅仅只是购物。它将成为一个集娱乐、文化和家庭活动于一体的目的地，为游客提供一个舒适、热情和谐的环境氛围。

In the north of culture center, neighcring Sunshine Paradise, the Galaxy Mall not only brings a high quality commercial experience to Tianjin, but also make the general site plan being good still better. The citizen square covers about 350,000 square meters, and contains functions of retail sale, hypermarket self-service shopping mall, dining and entertainment. This architeture coordinates around with other culture buildings in scale. The total height is limited to 30 meters. Each height from first floor to five floor is within 5.3-6 meters, and there are two layer basements on the level of -6.5m and -11m. Meanwhile, the consideration about the building's volume and materials is to form a graceful and harmonious among the grand theatre, the Sunshine Paradise and the commercial center. More than shopping, the commercial center produces a perfect commercial entertainment experience for tourists and resident of Tianjin. This center will be a destination for amusement, culture and family activities, providing a comfortable, passionate and melodious atmosphere

鸟瞰图
Aerial View

建筑体量由两个独特又相互联系的建筑形态构成：其一是呈立方体形状的零售商业；其二是位于南面倒映在水池和瀑布中的极具雕塑性和艺术性的水晶体。它们共同突出了“平衡与和谐”这两大设计主题。精致的水晶体背靠立方体，容纳着高档餐厅及特色店。或是作为家庭庆祝的场所，或是作为歌剧演出前后的进餐场所，或是提供便捷的午餐。水晶体表面的砾石状肌理将为人们的多种感官提供完整的体验。水晶砾石的立面采用节能的Low-e玻璃，多面体的立面由透明或半透明的玻璃组成。坐落在倒映水池之上，这些容纳餐饮空间的水晶体在对外展示商业中心内部消费活动场景的同时，也是眺望湖面、大剧院和远处博物馆建筑群的绝佳场所。

The Galaxy Mall is composed of two special shapes with mutural relationship: one is a cube regarded as the retail sale, the other is an artistic crystal with the sense of sculpture mirroring an image on pool and waterfall. Both of them reflect the design themes "peaceful and harmonious". Back to the delicate bended crystal, there are exclusive restaurants and specific stores, in which may hold family celebrations or be a dining place after performance in Grand Theatre, while the crystal stones bring multifarious sense experiences to people. The facades of the crystal stones are made of energy-saving Low-E glass, and that of the polyhedron are made of transparent or translucent glass strengthened. Lying On the water pool, these crystals not only are the landscape for internal shopping activities in the commercial center, but also provide a perfect place for overlooking the lake, the theatre and museum complex.

南立面 South View

北立面 North View

平 江 道
PINGJIANG Rd

水晶体夜景
Night View of Crystal Element Part

平面图 Plan

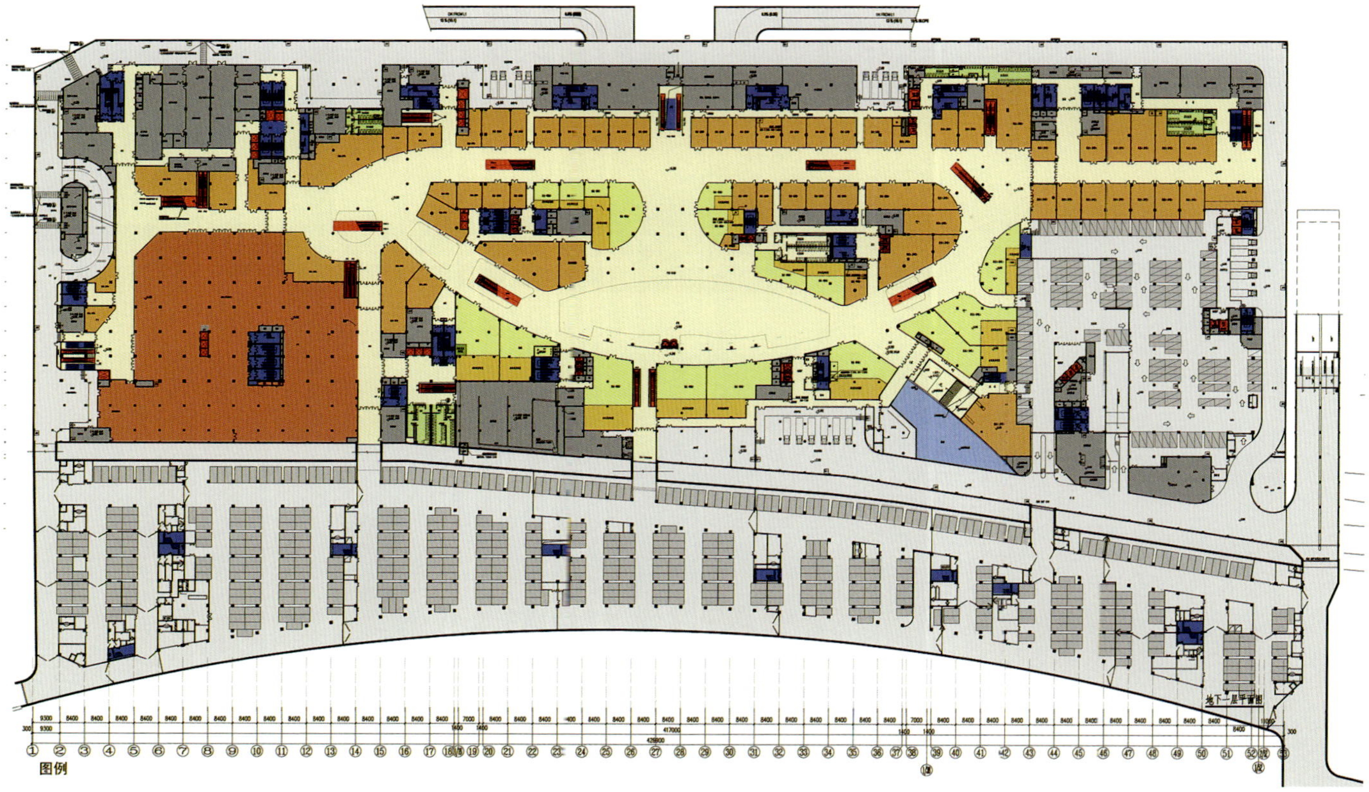

地下一层平面图 B1 Plan

图例 Legend

- 零售 commercial
- 百货店 department store
- 餐饮 restaurant
- 电影院 cinema
- 溜冰场 ice skating range
- 餐饮卫生间 restaurant toilet
- 卫生间 toilet
- 客用竖向交通 vertical transport
- 楼梯 staircase
- 公共区域 public space
- 设备用房 facility room
- 餐饮厨房 kitchen

首层平面图
F1 Plan

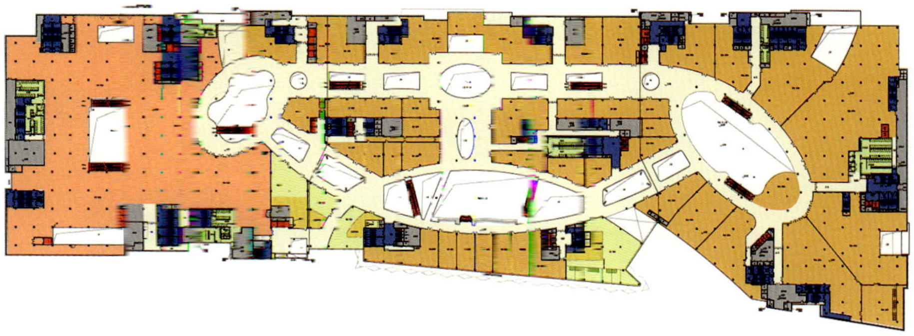

二层平面图
F2 Plan

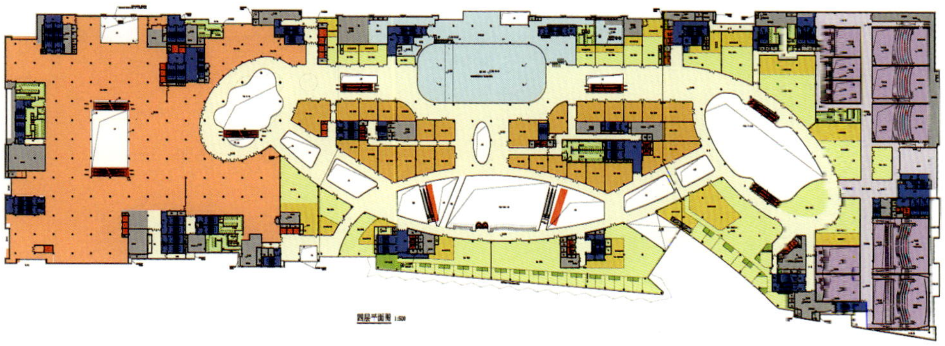

四层平面图
F4 Plan

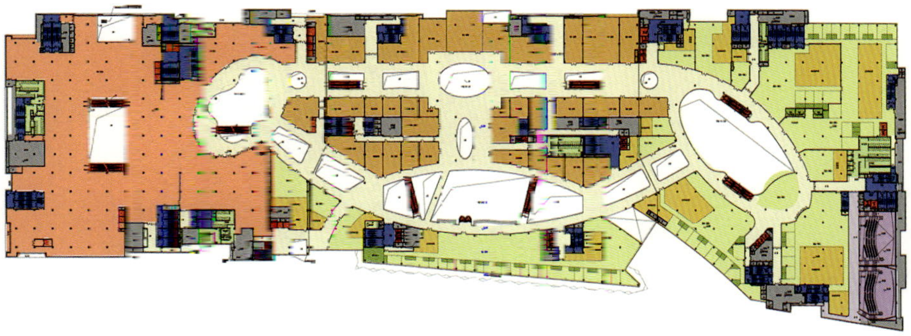

三层平面图
F3 Plan

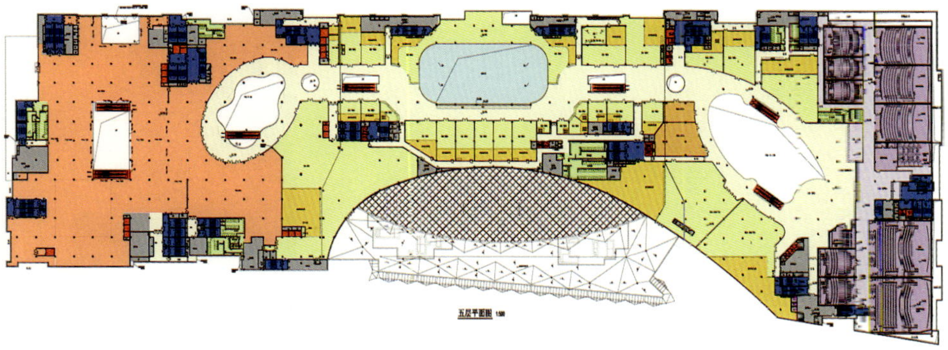

五层平面图
F5 Plan

水晶体
Crystal Elements

实景图
Outdoor Scene

水晶体
Crystal Elements

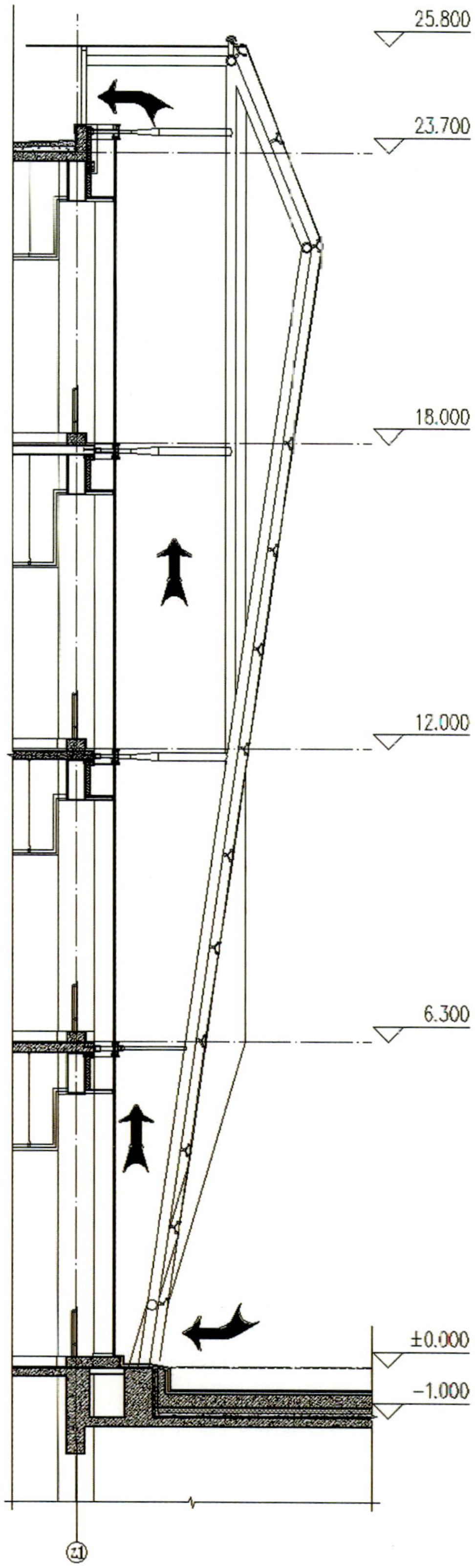

水晶体
Crystal Elements

幕墙
Curtain Wall

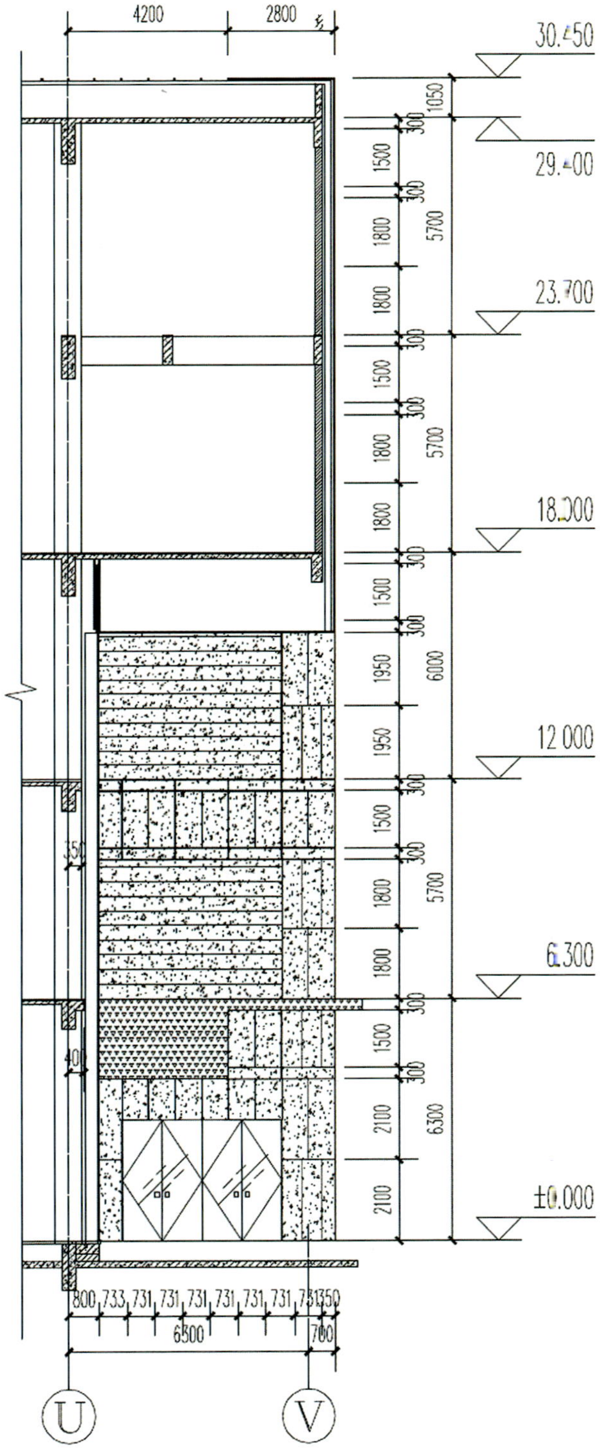

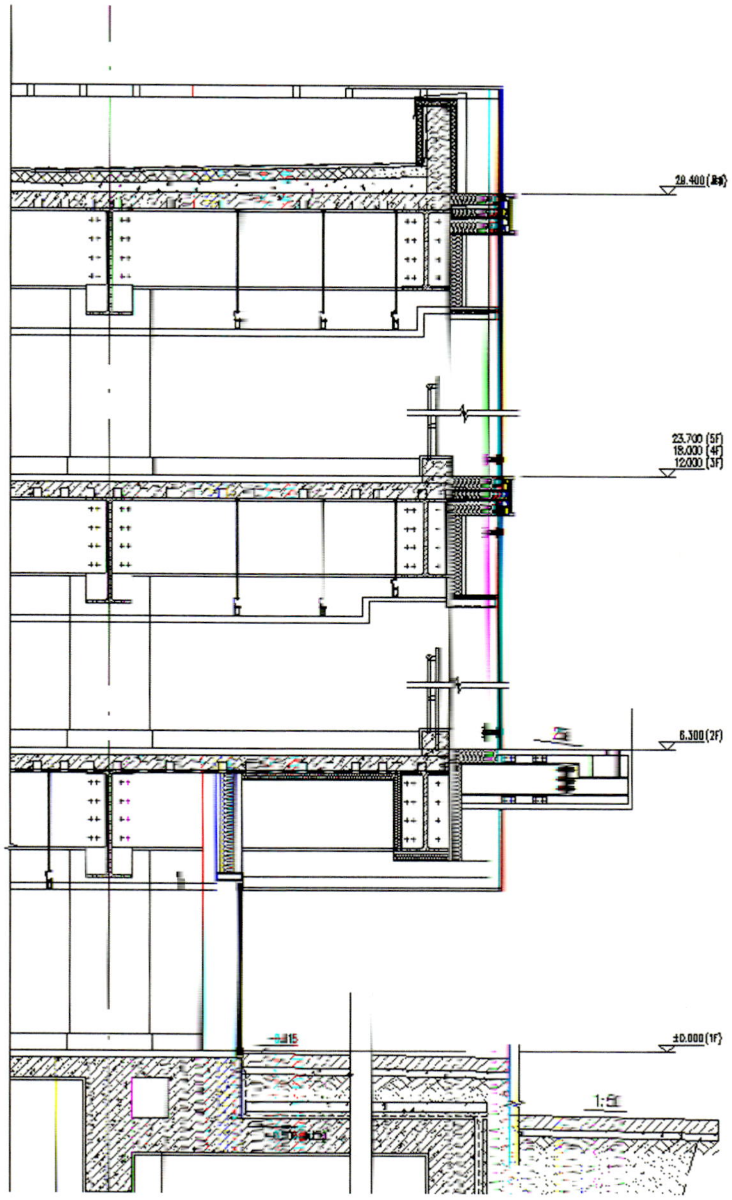
23.700 (5F)
18.000 (4F)
12.000 (3F)
6.300 (2F)
±0.000 (1F)

幕墙 Curtain Wall

幕墙局部
The Partial of Curtain Wall

幕墙外观
Exterior of Curtain Wall

太阳厅内景
Sun Hall Interior

建筑内装
Interior Decoration

月厅内景
Moon Hall Interior

建筑内装
Interior Decoration

星厅内景
Star Hall Interior

建筑内装
Interior Decoration

建筑内装
Interior Decoration

建筑内装
Interior Decoration

公共空间
Public Space

GALAXY
GALAXY
GALAXY

结构设计

本工程地下二层，地上五层。长410米，宽140米，地上设置了两道抗震缝，将整体结构分为三个分区，地下室不设缝。针对各个分区的特点，采用了不同的结构体系：建筑两端的A、C区，采用了钢筋混凝土框架—少量剪力墙结构，大跨楼层采用预应力混凝土楼面梁，共享空间周边超长细柱采用了钢骨混凝土柱；这样，既提供了开敞的商业空间，又保证了结构抗震的要求；在建筑中部的B区，考虑到大跨度无柱空间与水晶体的立面效果，采用钢框架—少量剪力墙结构，楼面承重构件采用钢梁和钢桁架，水晶体部分采用了空间管桁架；针对建筑物外檐的大跨度悬挑，采用了钢拉杆悬挂结构，使各层受力更加均匀、合理；屋面采用了网架、管桁架、实腹钢梁、钢拉索等不同的结构形式，以配合建筑专业形式多样、造型各异的艺术造型；同时，针对混合结构的特点，在钢骨混凝土、钢筋混凝土及钢结构节点处理上采取大量的创新做法。在地下室的设计中，为防止由于大体积混凝土收缩而产生收缩裂缝，在基础底板中施加了预应力，取得了很好的效果。由于地铁从基础下横穿，因此，基础设计采用了大跨度（约40米）桥式跨越，利用厚板完成了跨越和转换。

Structural Design

This project consist of two floors underground and five floors overground, totlely 410 meters in length and 140 meters in width.Two seismic joint are set on the building (not for basement level). As a result, the whole building is divided into three parts.Considering the structural characteristics in different parts, different structure forms are adopted :Section A and C located in the fringe of the building, and reinforced concrete frame with few shear wall structure system is suitable structural form. Prestressed concrete beam is used in large span floor, and steel reinforced concrete column is designed in this part of structure .This kind of structural system not only meets the demand making full use of the interspace of the commercial edifice, but also meets the limitation of seismic lateral displacement. Section B, as the central part of this building, requires space with large span, less columns, and crystal facade. Therefore, steel frame with few shearwall structure system is adopted and steel beams and truss are used as the sustaining member for floor system. For the crystal part of the building, space pipe truss is adopted in accordance to the curtain wall effect of the facade. In order to solve the problem of being surrounded by large span cantilever, a new creative measure using steel tension link suspension structure form is used so as to achieve the result of equal and reasonable bearing of load for each floor. Dormers vary in this project in terms of form and architectural design. Various of structural form, such as pipe truss spcace truss solid web steel beam , steel stayed cable used in this project to corperate the artistic effect of architecture and mixed structure is adopted in this project, quite a number flexible and creative technique measures is carried on during the designing process of node of steel reinforced concrete reinforced concrete and steel structure node. As a solution of the problem that may occour during construction process of its huge area and Large volume concrete foundation base plate, prestressing force is pressed in base plate, and get a very well effect. Because subway line should cross this area at the depth under its basement, so large bridge span(max span 40m) design method is used for the basement plate.

钢拉杆 The Steel Tie Rod

地球天窗结构 The Roof Structure of the Earth

钢筋混凝土梁与钢柱连接节点
The Reinforced Concrete Beam and the Steel Column Connected Node

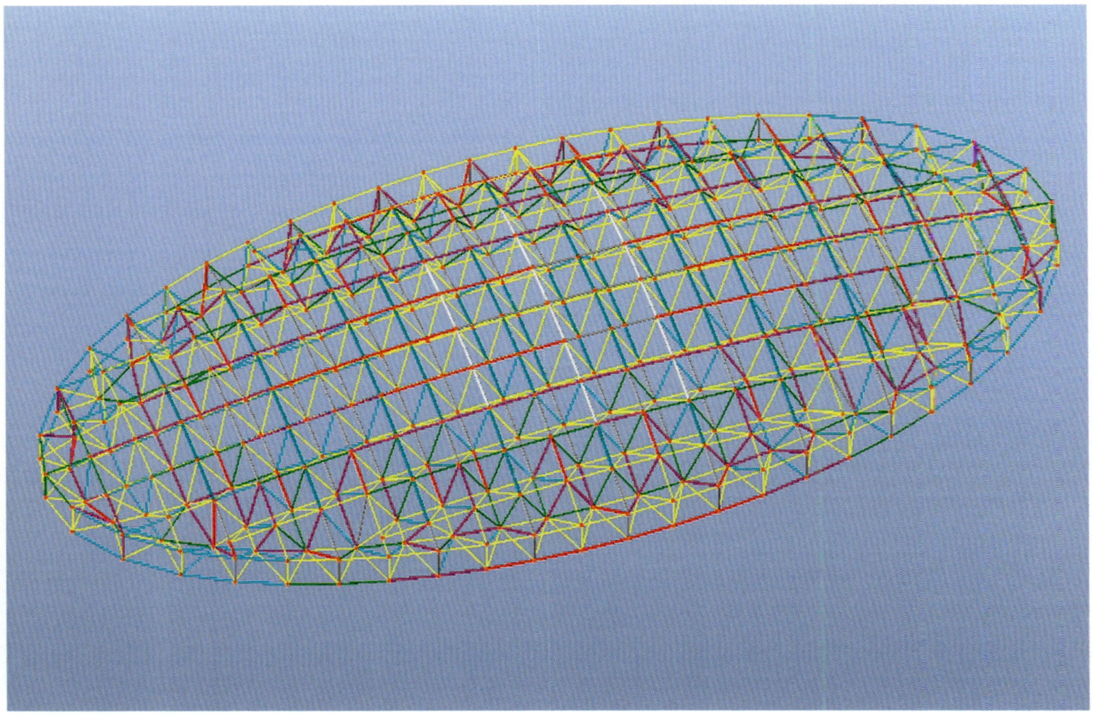

月亮天窗结构模型 The Moon Roof Structure Model

暖通空调设计

结合商业建筑人流变化大，店铺招商不确定性等特点，选取合理的室内设计参数。冷热源由文化中心能源中心提供，商业体于地下一层设置换热站。根据业态的功能分区划分为三个系统。空调冷热水系统均采用变流量系统。冷冻水温度为5.5℃/12.5℃，空调热水温度为45℃/37℃。冷冻水采用大温差设计，减小水泵选型，管道及管井，减少设备初投资及系统运行能耗。由于商业体舒适性要求较高，且进深较大，内外区分明。外区末端风机盘管采用4管制设计，内区末端风机盘管采用2管制设计，内外区所有的新风机组和空调机组全部采用4管制设计。末端设备可自由选择供热或供冷运行模式，各空调区域均能根据需求独立控制温度参数。在冬季内区冷负荷较小时，风机盘管可不开启制冷，通过新风机组送入经过预热的低温新风，消除内区冷负荷，保证商铺温度满足使用要求。小型商铺、餐饮店因独立经营，故将采用操作灵活的风机盘管的空调形式。另设新风机组提供室内人员所需新风量。部分新风机组设置热回收装置，以降低新风能耗。主力店、大型餐饮、走廊通道、中庭区域面积较大，功能、空调需求较为一致且负荷波动不大，此部分区域将采用全空气空调系统。

HVAC Design

Reasonable interior design parameters are selected according to the characteristic of commercial buildings crowd changes and uncertainty shops merchants. Hot and cold sources are provided by energy centers of the cultural center, the heat exchanger station is set in the commercial body basement. The body divided into three systems according to the format of the function. Air conditioning hot and cold water system is variable flow system. Chilling water temperature is 5.5 ℃ / 12.5 ℃, air conditioning hot water temperature is 45 ℃ / 37 ℃. The chilling water is with large temperature difference to reduce the pump, pipes and tube wells and reduce the device initial investment and system operation energy consumption.Due to the higher comfort requirements of the commercial body, and deep into the larger, inner and outer regions clearly. The end of the outer zone fan coil is 4-pipe design, the end of the inner zone fan coil is 2-pipe design, all the inside and outside new air handling units and air conditioning units are used in all 4-pipe design. End equipment are free to choose the heating or cooling mode of operation, the air-conditioned area can independently control temperature parameters according to demand. The zone cooling load is small in winter, fan coil may not open refrigeration, cryogenic preheat fresh air fed by the new air unit to eliminate the cooling load inner and ensure shop temperature to meet the requirements.Small shops, restaurants will use the air conditioning in the form of flexible operation of the fan coil due to independent operation. Another fresh air unit provides the indoor staff required fresh air. The fresh air unit with heat recovery reduces the energy consumption of new wind. The main shops, large dining, corridors, atrium area are large and consistent functional and air-conditioning demand, and small load fluctuations, this part of the region will adopt the whole air air-conditioning

给排水设计

给水和中水系统采用分区供水方式，首层和地下室利用市政水压直接供水，二-五层为变频泵加压供水。供水管道在各层呈环状布置，方便商铺精修时用水点的接驳。引入中水系统冲厕和绿化。各喷泉水景设置了独立的循环水处理系统。其中大地中庭设有大型数字水帘式喷泉，水帘从四层逐层跌落，并伴有商家的标志性图案，极为壮观。主入口设置了跌水式瀑布。室外设有镜面景观水池，利用错峰蓄热的方法对池水进行了加热，达到冬不结冰夏清澈、四季如画的景观效果。餐饮含油污水采用室外隔油池和室内一体化设备相结合的处理方法。屋面设置了虹吸式雨水排水系统，减少了室内雨水立管的数量和雨水干管占据的空间高度。消防系统按照防火规范和消防性能化报告的要求，设置了消火栓系统、喷淋系统、大空间智能型主动喷水灭火和气体灭火系统。为保护店铺与走道分隔的钢化玻璃，设置了窗式喷淋系统，用国产快速响应侧喷式喷头取代了昂贵的进口窗式喷头，经火灾试验检验，能够达到阻火灭火的功效。为了解决上千吨的消防水池水质污染的问题，在消防泵房设置了屋面光伏发电电池板的冲洗水泵，从消防水池取水，定期冲洗电池板，使池水不断更新流动，水质得到了良好的改善，并在冲洗水泵的吸水管采取了措施，保证水池内的消防用水不被动用。

大地厅的数字式水帘和跳泉
Digital Curtain-Style Fountain in Earth Hall

Water Supply and Drainage Design

Water supply and recycle are zoning supply, the municipal pressure water supply directly are used in the first floor and basement, two - five layers are using the variable frequency pump pressurized water. The water supply pipes in layers are arranged in a ring which is convenient in connection when shops decoration. Recycled water system is added for flushing and gardening. Independent circulation water system was set in fountain. There is a large digital curtain-style fountain in the courtyard, curtain drops from four layers, accompanied by businessmen iconic pattern, which is spectacular. Waterfall is set in the main entrance. Mirror landscape pool is set outside, peak heat storage method is used for water heating that there is no ice in winter, clear in summer and picturesque landscape effects. The combination of outdoor grease traps and indoor integration equipment are used for catering oily wastewater. Siphon rainwater drainage system is set in the roof for reducing the quantity of rainwater vertical pipe and the height space the rainwater main pipe occupied. Fire protection system contains the fire hydrant system, sprinkler system, large space intelligent active sprinkler and gas fire extinguishing system in accordance with the requirements of fire safety regulations and fire performance report. Window sprinkler systems is set with the domestic fast response side spray nozzle to replace expensive imported windows sprinklers for the protection of tempered glass separating the stores and aisles. In order to solve the problem of thousands of tons of fire pond water pollution, roof photovoltaic power generation solar panels flushing water pump is set in fire pump. Fire pond water regularly flush battery plate, so that water constantly updated flow, water quality has been good improvement. Take measures in flushing water pump suction pipe to ensure that the fire water in the pool not utilized.

屋顶光伏发电建筑一体化及系统集成设计

本项目位于北纬38° 34', 东经116° 43', 水平面年辐射量为1657kWh/m²，日平均峰值日照小时数4.0h,年平均温度12.3℃；极端最低温度-18.3℃；极端最高温度39.7℃。该工程建成后直流额定发电功率为300.132kW。在屋顶共铺装7146块42W非晶硅薄膜电池，总铺装面积为 7370m²。该发电系统主要由光伏组件、直流集线箱、交流并网柜、逆变器，防逆流和监控系统等组成。通过系统组件—逆变器内部的最大功率点跟踪技术和主动式防孤岛保护方案，来保证太阳能发电逆变输出最大电能和用电安全。

在设计实施过程中，充分利用建筑屋面架构的金属格栅条件，平铺光伏系统组件——电池板，解决了电池板维护方面通风散热问题。其次，为保证系统发电效率，在屋面上设置十多处自来水上水点，定期清洁板面尘埃。考虑储能原件—电池组寿命及运行维护特点、并结合经济造价比较，确定了即时发电即时消耗,不储能方案。本系统采用多点三相并网（内部电网）不上网（市政电网），专电专用模式，为建筑内车库照明负载提供清洁的绿色电力。该发电系统建成后，年发电量可达35.94万度，可省燃油93432.04升或节省标准煤129367.44千克，相当于少排放至大气358275.938千克二氧化碳，4240.3772千克二氧化硫，1545.2222千克氮氧化物。同时减少因火力发电产生的97744.288千克粉尘，节约1437416升净水。

Rooftop Photovoltaic Building Integration and System Integration Design

The body-Rooftop photovoltaic building integration and system integration project, located in north latitude 38° 34', longitude 116° 43', horizontal plane annual radiation 1657kWh/m²,daily average peak sunshine hours of 4.0h, the annual average temperature of 12.3℃; extreme minimum temperature of -18.3℃; extreme maximum temperature of 39.7℃. After the project is completed, the DC rated power 300.132kW. 7146 42W amorphous silicon thin film battery works on the roof of pavement, pavement area of 7370m², bracket tiled installation. The power generation system includes photovoltaic modules, DC set line box, communication and network cabinet, inverter, backflow prevention and monitoring system, etc. To ensure that the maximum power output of solar power inverter and electrical safety by system components-inverter within the maximum power point tracking technology and active anti-islanding protection program.

In the design and implementation process, first of all, give full maintenance building the metal grille conditions of the roof structure, tile photovoltaic system components - solar panels, to solve the battery ventilation problems. Second, to ensure the power generation efficiency of the system, more than 10 tap water supply points are set on the roof, periodically clean the plate surface dust. Third, consider the storage originals - life and run maintenance characteristics of the battery pack, combined with the economic cost comparison, determine the immediate consumption of the instant generation of photovoltaic power generation system, not stored energy program.

The photovoltaic power generation system uses multi-point three-phase network (internal grid) ,unconnected to the power grid (municipal grid), the special telegram special mode. The photovoltaic power generation system and the mains system provide clean green electricity for underground garage lighting. After the completion of the power generation system, Annual generation capacity is up to 359,400 degrees, saving fuel 93432.04 liters or savings of 129,367.44 kg of standard coal, equivalent to less 358,275.938 kilograms of carbon dioxide emissions to the atmosphere, 4240.3772 kilograms of sulfur dioxide, 1545.2222 kilograms of nitrogen oxides. While reducing 97,744.288 kilograms of dust due to thermal power generated and saving 1,437,416 liters pure water.

屋顶太阳能电池板 Roof Solar Panels

局部放大1
Detail of Solar Panel 1

局部放大2
Detail of Solar Panel 2

交通枢纽工程
Transportation Hub Project

上海市城市建设设计研究院
Shanghai Urban Construction Design Research Institute

地下交通主枢纽工程，总面积为14.48万平方米，主要包括结合5号线文化中心站实施的地下交通枢纽配套设施，同步实施与其换乘的地下三层轨道交通站体土建工程，建筑面积为54414平方米，其中近期利用地下一层16500平方米暂作为机动车停车场，综合配套服务用房，下穿越秀路连廊，地下机动车停车场。

①综合配套服务用房，建筑面积为 65774平方米包括地下商业部分建筑面积56724平方米，夹层非机动车库建筑面积2186平方米，西区能源中心建筑面积6864平方米 .

②下穿越秀路连廊, 建筑面积为2169平方米
通道最宽处约36米, 最窄处约为33.6米, 西侧与地下交通枢纽连接, 向东至越秀路红线以东3.2米 总建筑面积约为2328平方米。整条通道净高不低于3.5米。

The construction area of underground transportation hub project is 144,800 m², including the underground transportation hub supporting facilities area which combines with the cultural center station of subway line 5. Civil engineering work of the three-layer railway station and culture center station of subway line 5 have been implemented simultaneously. The total construction area is 54,414 m². The 16,500 m2 basement level one will be used as parking lot and comprehensive services space will connect underground parts of Yuexiu Road.

①Construction area of comprehensive services facility is 65,774 m².
Construction area of underground commercial is 56,724 m²; mezzanine non-motor vehicle parking is 2186 m², and Western Energy Center is 6,864 m².

②The corridor area of construction below Yuexiu Road is 2,169 m²
The widest space is around 36 m; the narrowest space is around 33.6 m. The northern corridor connects to the transportation hub and the eastern corridor extends 3.2m beyond the red line of Yuexiu Road. The overall floorage is approximately 2,328 m². The net height is over 3.5 m.

角广场
Corner Sunken Plaza

③地下机动车停车场，建筑面积为15966平方米
地下机动车停车场建筑面积为1.6万平方米，车位数295辆，为地下一层，底板标高为-9.000m，顶板标高为-2.600m，上侧的覆土为2.0m,车库层高为6.4m,吊顶下净高不小于3.5m。即可以满足小型车的停车需求，也可以满足中型车的停车需求。（以上标高均采用相对标高， 0.00m相对于08大沽高程3.20m。）

④地铁5、6号线换乘通道，建筑面积为2390平方米
5、6号线换乘通道面积约为2390平方米，为地下二层，底板完成面标高为-16.200m，顶板完成面标高为-9.300m，上侧为下沉广场，换乘通道层高为6.9m，净高不小于3.5m.（以上标高均采用相对标高， 0.00m相对于08大沽高程3.20m。）

③The construction area of underground parking lot is 15,966 m².
The construction area of underground parking space can hold around 295 vehicles. The height of basement floor is -9.000 m and the height of basement ceiling is -2.600 m. The earth is 2.0m thick above the basement and the height is 6.4m. The net height is over 3.5m which is suit for both small car and middle car. (The ground elevation is relative height which the 0.00 m is 3.20m higher than 08 standard elevation)

④The construction area of transfer tunnel between Line 5 and Line 6 is 2,390 m²
The construction area of transfer tunnel between Line 5 and Line 6 is 2,390 m², which located at B2 floor. The height of basement floor is -16.200 m and the height of basement ceiling is -9.300 m. The transfer tunnel is below sunken plaza which is 6.9 m high and the net height is over 3.5 m. (The ground elevation is relative which the 0.00 m is 3.20m higher than 08 standard elevation)

公交候车廊
Passenger Waiting Area

公交候车室
The Waiting Room

本工程结合地面广场景观主要设置三个大型下沉广场，通过下沉式广场使地面广场的开放性空间与地下广场的配置具有相关性，保持地面广场的完整性，同时结合地面绿化设置开敞的地下连通道，解决了地下空间的防灾问题并将自然环境引入地下。同时考虑北方天气因素，下沉广场局部加顶盖，既方便市民出行也降低日常维护费用。

Three large sunken plazas areas combined with the floor-level plaza keep the integrity. Open lane set to connect the sunken plaza with the landscape which solves the disaster prevention by using the underground space. Considering weather effects in northern China, part of sunken plaza constructs will have a cover which reduces the maintenance cost and enhance convenience to citizens.

花园下沉广场
Garden Sunken Plaza

中央下沉广场 1
Central Sunken Plaza 1

中央下沉广场 2
Central Sunken Plaza 2

4线换乘交通枢纽

由于文化中心地下空间内部及周边有4条地铁线路穿过，包括M5号线、M6号线、M10号线及Z1线，在文化中心形成了一个地铁换乘的枢纽站，通过规划设计，统一考虑了4条地铁线之间如何方便快捷的换乘，如果将地铁人流引入其周边的地下商业空间，社会车辆与地铁之间如何换乘，公共交通与地铁之间如何换乘，自行车与地铁之间如何换乘等问题，通过统一规划设计，合理安排布局，实现了地铁与地铁、自行车、公交车、社会车、商业人流之间真正意义上的便捷换乘，加强了文化中心的交通枢纽功能。

4-Line Interchange Station

The cultural center interchange station of 4 subway lines includes Line M5, Line M6, Line M10 and Line Z1. The design of the interchange station focuses on convenient transfer among the 4 lines. Circulations of people are lead to underground business area. The design solves the of transfers between cars and subway, transfers between public transport and subway, and transfers between bicycle and subway. The design enhances convenient transfer among subway, bicycle, vehicle and people. The interchange station makes cultural center more functional in transportation.

矩形天窗
Rectangular Skylight

主通道天窗
Skylight of the Main Pathway

大体量平层商业

越秀路以西地下空间建筑面积约10万平方米，包括约7万多平方米的地下商业空间，整个地下商业空间基本为地下一层，以平铺的方式安排布局，长轴方向长度约为450米，商业定位通过策划公司的调研及建议，确定为中低档，商业功能布局包括一个大型超市以及部分餐饮及散铺相结合，各商业功能通过宽度为13.5米的商业主通道和宽度为9米的商业次通道组织。

Flat Underground Commercial

The underground construction area is around 100,000 m² to the west of Yuexiu Road including 70,000 m² commercial. The entire B1 floor is designated for commercial. The horizontal axis is approximately 450 meters. The commercial type focuses on low and middle level as the result of strategic planning including a supermarket and some restaurants combined a pathway. Shops area along the 13.5 meters width main pathway with a 9 meters width secondary pathway.

机动车库
Underground Garage

非机动车库
Non Motorized Parking Lot

机动车库
Underground Garage

平面图 Plan

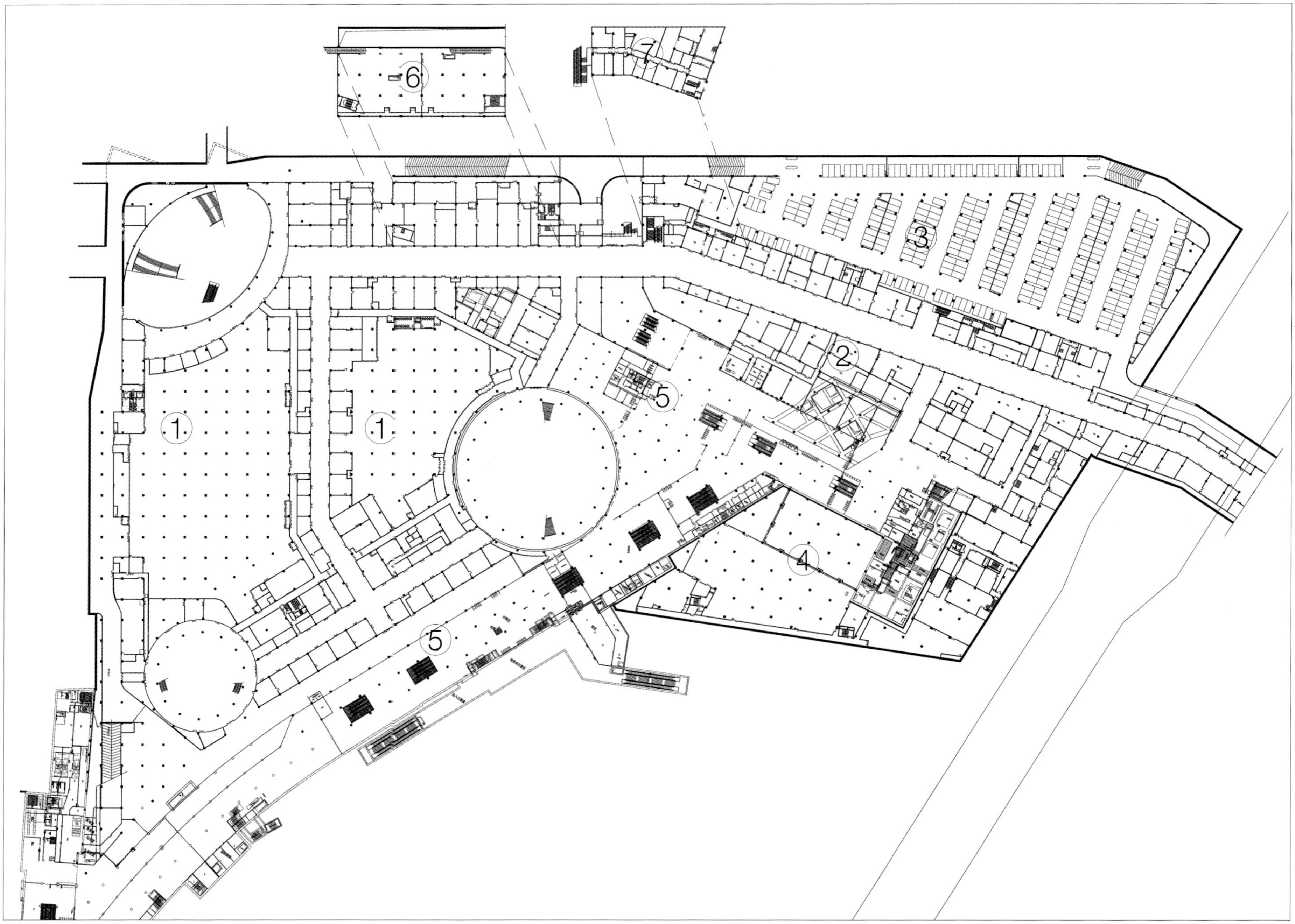

地下一层平面图
Basement Level One Plan

① 商业区 commercial area
② 餐饮区 dining area
③ 机动车库 underground garage
④ 能源中心 energy center
⑤ 地铁 subway
⑥ 非机动车库 non motorized parking lot
⑦ 公交办公 office for public transportation

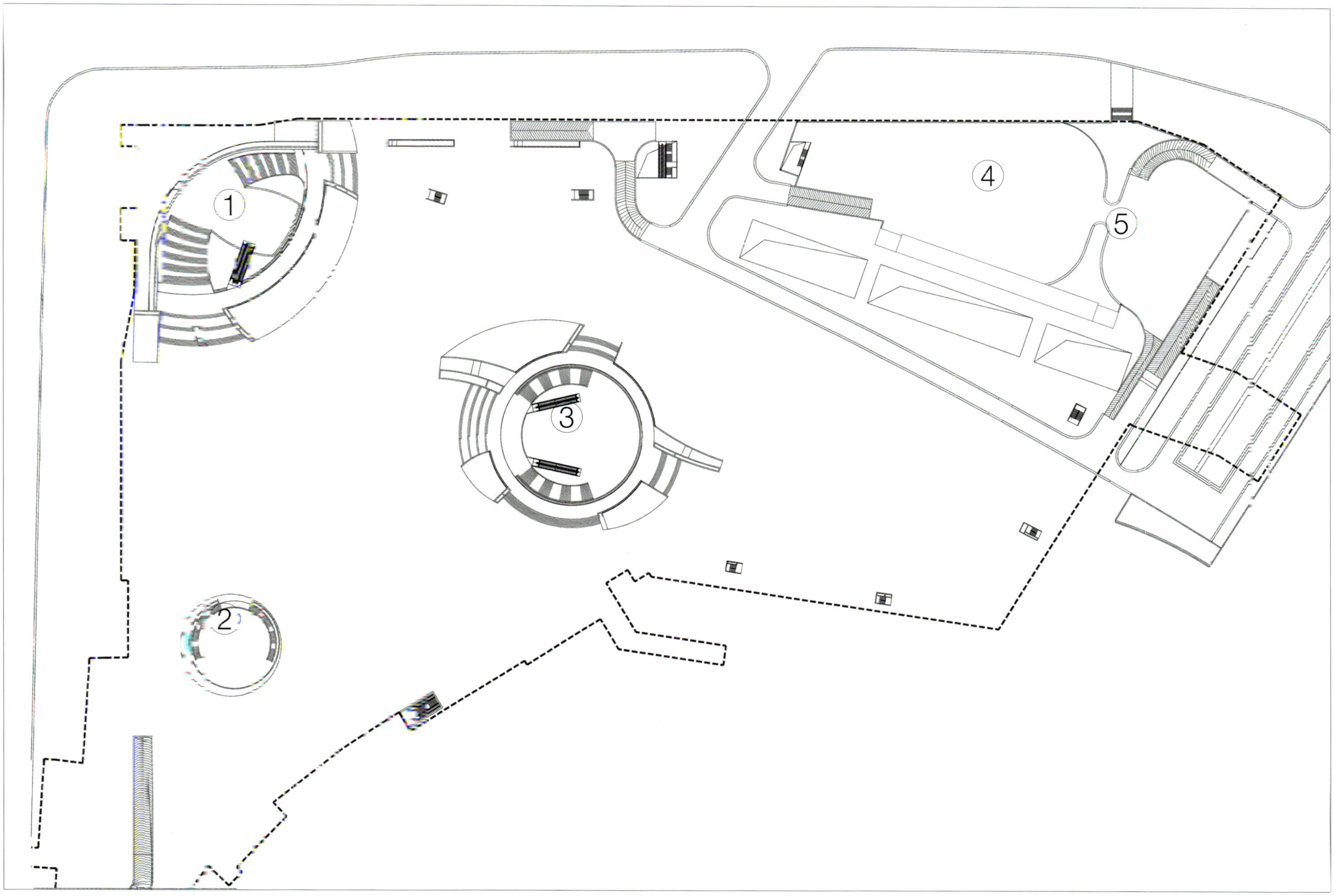

地下一层平面图
Basement Level One Plan

① 角广场 corner sunken plaza
② 花园广场 garden sunken plaza
③ 中央广场 central sunken plaza
④ 公交首末站 bus terminal
⑤ 大巴停车场 bus parking lot

文化中心能源站
Cultural Center Centralized Energy Station

天津市建筑设计院
Tianjin Architecture Design Institute

文化中心集中能源站为近100万平方米的文化、科技、商业综合体建筑的采暖与空调系统提供热源与冷源，实现了可再生能源在建筑上的规模化应用，体现了可持续发展的规划设计理念，具有示范性。由于利用了浅层地热能和实现了空调排热的跨季节蓄存，显著降低了供冷供热的化石能源消耗和污染物的排放，实现了“绿色”供能。地源热泵的采用使向大气排放的空调废热负荷减少80%，冷却塔装机容量减少了75%，既降低了城市热岛效应，又避免了对建筑第五立面的景观影响，同时还可以节约大量的水资源，而蓄冰技术的应用则降低了夏季城市电网的负担，实现了城市电力的削峰填谷。

能源站所采用的能源形式与传统的能源形式相比，每年可节约标准煤8873吨，减少CO_2排放23249吨、减少SO_2排放213吨、减少NOX排放81吨，节水率为55.43%，节能率为36.02%。

摄影：刘东
Photographer：Liu Dong

Tianjin Cultural Center centralized energy station provides cooling and heating source to meet the needs of cooling in summer and heating in winter for 1 million square meters buildings of cultural, technical and commercial types, realizing the renewable energy was applied on heating and cooling of buildings on a large scale, and it embodied Tianjin Cultural Center's sustainable development planning and design philosophy, thus playing an exemplary role. Owing to adopting shallow geothermal resource and achieving the over season heat storage of the waste heat discharged by air conditioning system, the consumption of fossil energy and emission of pollutants were reduced significantly, thus the heating and cooling energy is green. Similarly profiting from the ground source heat pump system, the waste heat of air conditioning which originally should be discharged into outdoor air have been reduced by 80%, as well as the installed capacity of cooling tower was decreased by 75%, Hence the urban heat island effect was reduced, but also the bad impression on building-roof was avoided because of cooling tower was dispelled, simultaneously, saving a lot of water resources, By the ice storage technology, the burden of urban power network can be reduced and the peak load shifting of the city power system can be achieved also.

Compared with the traditional energy form of heating and cooling, the sustainable central energy system of Tianjin Cultural Center can cut down annual consumption of CE by 8,873 tons, the CO_2 emission by 23249 tons, the emission of SO_2 and NOX emissions by 213 tons and 81 tons separately. Simultaneously, the rate of water saving is 55.43%, and the rate of energy conservation is 36.02%.

西区能源站 Western energy station

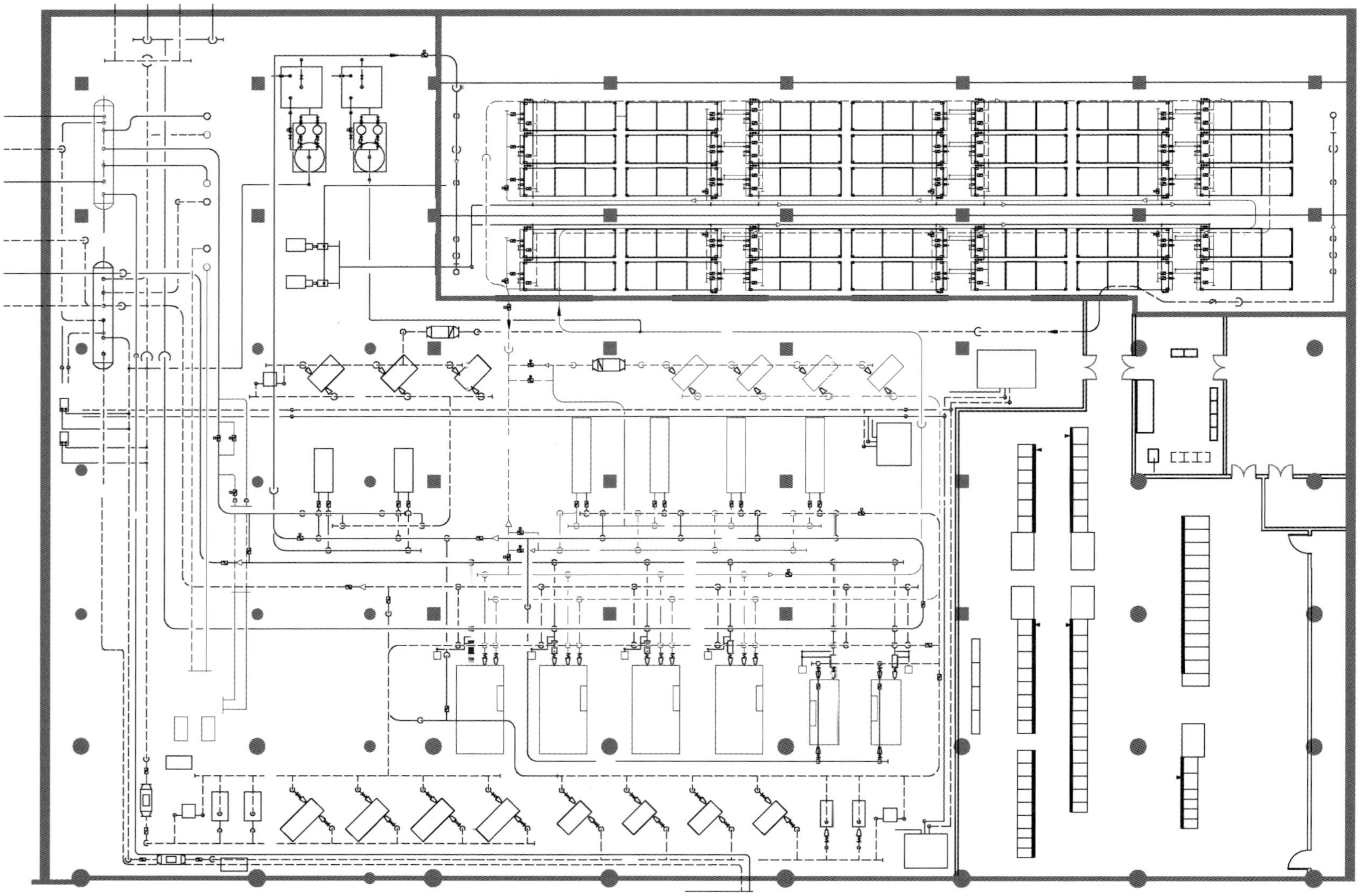

能源站平面图
Layout of Energy Station

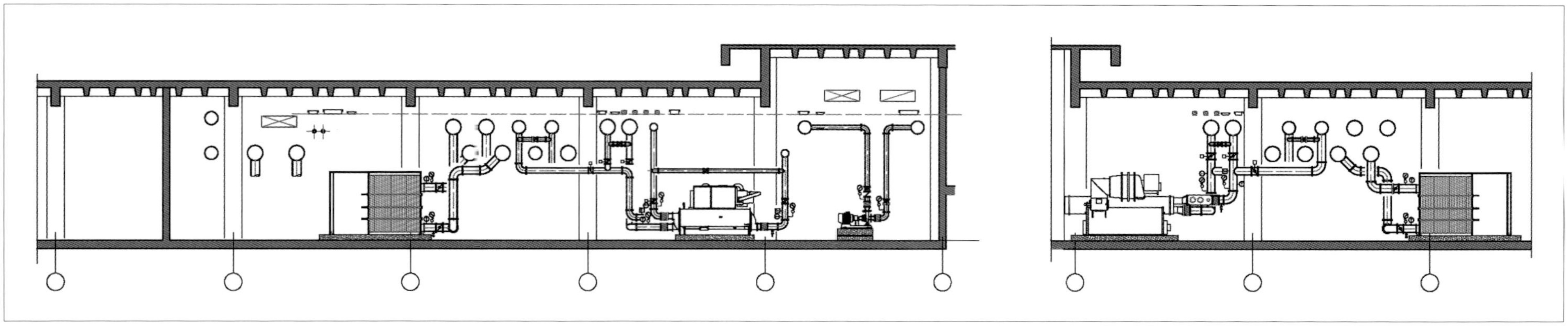

剖面图 Section

蓄冰盘管 Ice Storage Coil

蓄冰槽 Ice Storage Slot

配电室 Distribution Room

变电室 Substation

文化中心区域能源系统针对不同的业态特征及管理权属，结合地下空间设置西区、南区、北区三座集中能源站，可实现联网运行，总供冷负荷85875kW，总供热负荷52116kW。

能源站工艺系统形式为设有冷、热调峰复合以冰蓄冷技术的三工况水/地源热泵系统，其中冷调峰为下沉安装的鼓风式冷却塔，最大限度减少了对景观的影响；热调峰为热电厂提供的市政热网，调峰比例25%；冰蓄冷采用主机上游、串联布置、钢盘管冰槽蓄冰、外融冰取热方式；地源热泵分为两种形式：西区采用250m/400m深16对浅层地下水水源热泵系统，水源井布置于生态岛；南区、北区为垂直土壤埋管地源热泵系统，3789孔埋管位于景观湖底，充分利用了自然资源并与景观有机结合。

According to the ownership and the function of different buildings, the sustainable central energy system of Tianjin Cultural Center involved three centralized energy stations named as western, southern and northern station which are located in the underground space, and the outdoor pipe networks were connected each other. The systems gross cooling load is 85875 kW and gross heating load is 52116 kW.

The central energy station adopts compound three operating conditions energy system combined water/ground source heat pump, ice storage, peak-load cooling and heating source. The peak-load cooling source is the forced draft cooling tower installed in a sink space on the ground, minimizing the impact on the landscape, and the peak-load heating source comes from municipal heating network which is offered by the thermal power plants, the peaking ratio is 25%. The form of ice storage refrigerating system is chiller upstream and arranging the steel coil ice tank in series, and external melting. There are two types of ground-source heat pump systems employed, the western station adopts shallow groundwater system with 16 pairs of heat source well which are 250/400 meters deep and drilled in ecological island; the southern and northern station employed vertical ground heat exchanger system with 3789 bore-hole under the bottom of landscape-lake, resulting in both taking full use of shallow geothermal resources and achieving a beautiful landscape-lake.

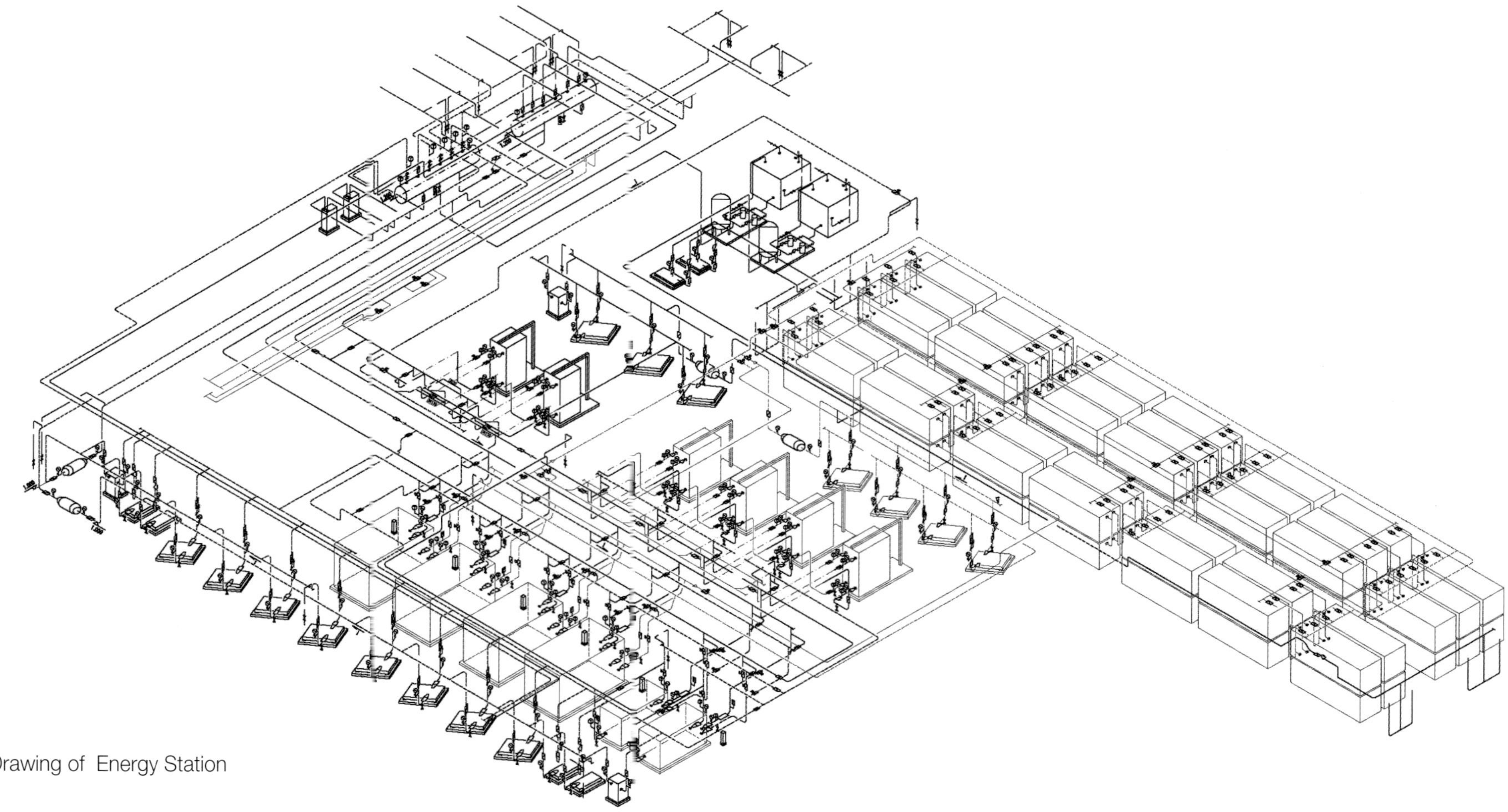

能源站系统图
The System Drawing of Energy Station

南区能源站
South Energy Station

集中能源站采用了基于神经网络负荷预测控制算法的前馈型能源管理与自动控制系统，通过实时负荷预测与追踪，实现“按需供能”，最大限度提高能源系统效率，节约能源。同时，通过对垂直土壤埋管换热器区域的土壤温度、水源井参数、能源站工艺流程以及用户侧换热站参数的实时在线监控与管理，保证集中能源系统的安全、长期、高效运行。能源站之间采用基于TCP/IP的星型网络结构，实现互联互通。

Centralized energy stations employed the feed forward type energy management and automatic control system which is based on the neural network load forecasting control algorithm. This system can realize on-demand energy supplying, improve the efficiency of energy system and achieve energy conservation as far as possible by real-time forecasting and tracking of heating and cooling load. Meanwhile it can also ensure that the centralized energy system has a long-term safety and efficient operation, depending on the real-time online monitoring and management of the vertical ground heat exchanger area's soil temperature, processing cycle of energy stations, parameters of heat source wells and ETS. The star network structure based on TCP/IP is adopted among each energy station to achieve interconnection and interoperability.

于景观湖底进行土壤源热泵系统 3789 口垂直埋管换热器的敷设
3789 bore-hole under the Bottom of Landscape-Lake

湖底埋管
The Buried Pipes Located in the Bottom of Lake

管控中心
Management and Control Center

天津市城市规划设计研究院
Tianjin Urban Planning & Design Institute

管控中心北侧临近大剧院与科技馆，南侧紧临青年京剧团，是文化中心东南角街坊内最后一块“拼图”。后来者的角色，以及服务性建筑的特点，决定了管控中心需要“低调”亮相，与街坊内的青年京剧团、中华剧院组成整体协调的建筑群，共同融入到文化中心。

建筑体量呼应广东路弧形道路红线，采用与青年京剧团相似的体量组合方式，以两个条状体块错位拼合成反“Z”形组合。从相邻的青年京剧团中提取红墙、坡顶、深灰色钢架等建筑元素，通过更加简洁现代的立面划分方式，形成明快且含蓄的建筑特色。相近的建筑元素与东南角街坊内原有建筑形成对话，简洁的风格在折衷的京剧团建筑与现代的大剧院、科技馆之间形成过渡。

管控中心主要功能是为保障文化中心建成后正常运转而设置的。包括控制中心、执法大队、派出所、科技馆辅助用房，以及中心湖面大型喷泉设备用房、配套地下停车库、设备用房。在使用单位较多，功能复杂的条件下，反“Z”形的体量组合在用地内形成了供各单位使用的相对独立的入口空间，减少了相互干扰。

摄影：陆伟伟　田垠　刘东
Photographer: Lu weiwei,Tian yin,Liu Dong

总平面　Master Plan

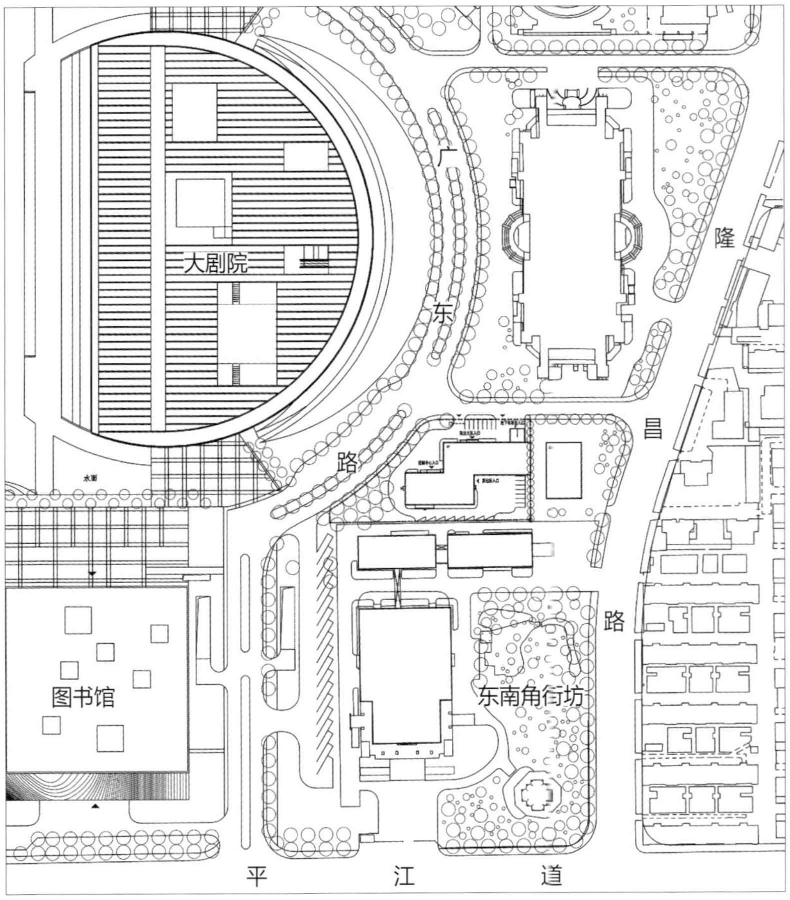

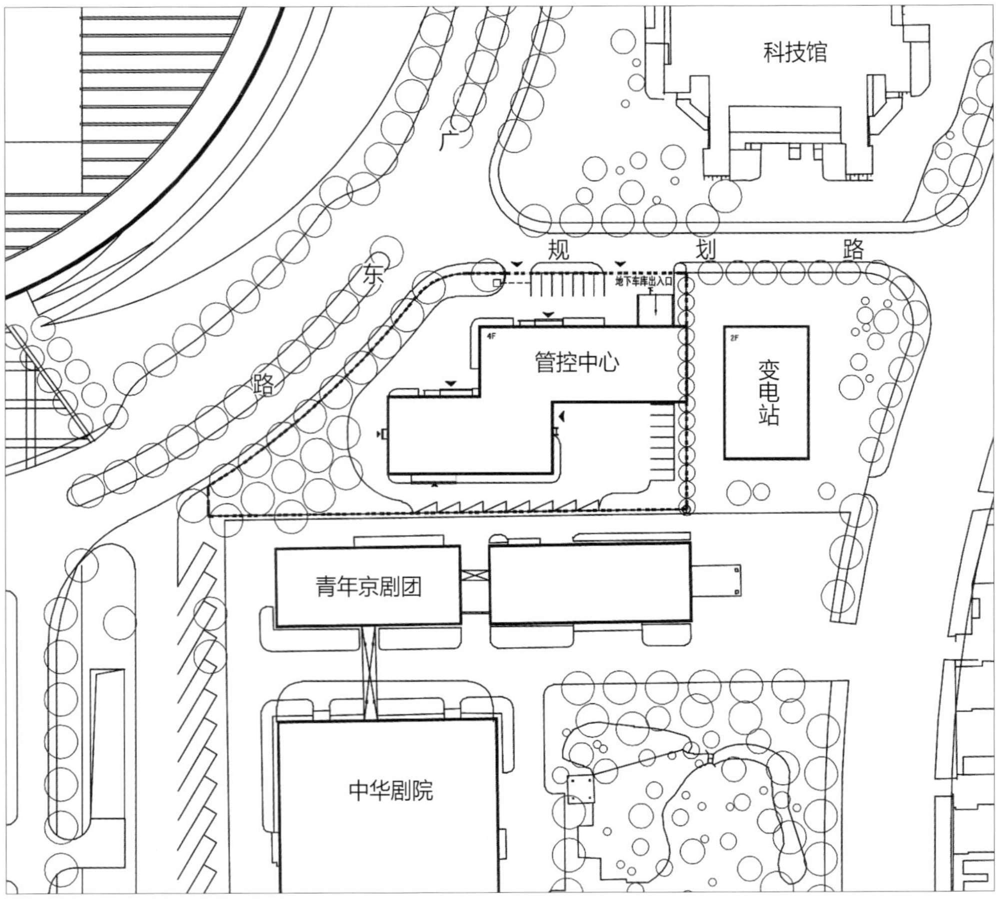

Management and control center is close to grand theater and technology center. Management and control center is the last piece of jigsaw in the northeast site of cultural center and close to the Youth Peking Opera Troupe House to the south. Since the management and control center is designed in the last phase of cultural center and as a service building, it performs a low-key position. The design of management and control center cooperates harmoniously with surrounding buildings such as Youth Peking Opera Troupe House and Zhonghua Theater.

Building volume of management and control center correspondes to the arc of Guangdong Road; the two block buildings make up a Z shape by adopting similar building volume with Youth Peking Opera Troupe House. By extracting red wall, slope ceiling and deep grey steel structure elements from Youth Peking Opera Troupe House and combining concise modern façade, management and control center is both implicative and succinct. Similar architecture elements correspond to buildings in southeast corner. Concise building style transforms gradually and become a tradeoff between Youth Peking Opera House, Grand Theater and Science and Technology Center.

The main function of management and control center is to guarantee the normal operation of the cultural center including control center, local policy station, law enforcement group and science technology auxiliary building. The control center also includes equipment auxiliary building of the musical fountain of the central lake, underground parking and other auxiliary buildings. The Z-shaped building form allows individual use and non-interruption use with many companies and functions in it.

鸟瞰 Aerial View

北侧实景图 North View

西北角实景图 West North

北侧实景图 North View

北侧实景图 North View

3

环境设计篇

Part 3: Environmental Design

景观设计 Landscape Design

德国戴水道设计公司 Atelier Dreiseitl GmbH
ATLAS（中国）规划· 设计 ATLAS (CHINA) Planning & Design
天津市城市规划设计研究院 Tianjin Urban Planning & Design Institute
天津市建筑设计院 Tianjin Architecture Design Institute
天津市园林规划设计院 Tianjin Institute of Landscape Planning & Design

天津文化中心的园林景观犹如声音渐起的交响乐与轻柔的横笛独奏所形成的清晰对比，与环绕她的周边建筑形成优美和谐的城市画卷，吸引着四方游客。伴随国际设计竞赛后的近三年的紧张工作，一个由景观设计师、规划师、和建筑师组成的设计团队完成了文化中心的设计创作和施工，从建筑到园林，由室内到室外，一个个文化场所通过中心湖-这个艺术化的水体有机地结合起来。中心湖位于项目的核心位置，其北侧的城市步行大道一直延伸至文化中心的视觉焦点-天津大剧院。湖的四周，一系列的生态花园、广场、平台向游客讲述有关天津文化和艺术的故事，并且引导人们来到自然、历史博物馆、购物中心和青少年活动中心等文化建筑。如此集中的在城区内展示文化建筑群，同时又能在如此短暂的时间内高质量完成，在国际上实属少见。此项目的成功承载着中国政府及设计师对天津民众的承诺及责任，同时戴水道设计团队为能参与此项目、共同见证中国城市发展而感到自豪。

总体设计从文化公园本身出发。场地共计约90公顷，中心湖面10公顷。700米长的步行大道贯穿东西，在大剧院面前150米宽的滨水台阶上形成最优越的观景点。在这个大尺度的构架下，沿着人行的流线布置了一系列人性化尺度的场所，给人们提供静谧的休闲空间。总体设计的目标是围绕现状自然历史博物馆，打造一个具有国际水准的文化区。深具人文情怀的景观设计，在倡导强烈文化诉求的同时兼顾了市民休闲运动的日常需求，设计了滚轴区、野餐场地以及家庭活动的空间。因此，文化中心既着眼于国际，又立足于天津本地人民的需求。

文化中心整体上贯穿着动态的城市水景主题，突出自然元素的中心湖是整个景观设计的核心。就像一幅美术馆里的油画，它反映着人们的生活。湖四周的边界设计紧密结合周边的场所环境。北侧边界是优雅的城市步行大道，它连接城市与水面，具有雕塑感的波浪纹理和灯光的反射进一步强调这种联系。南侧的美术馆平台潜藏于水面之下，提供独具特色的室外展览空间。西侧生态公园的水岸是自然的，在形态上更加灵活多变，有芦苇地、木甲板、生态净水群落、和引人入胜的内部水湾。东侧的大剧院平台作为城市舞台，让人们能够拾阶而下，亲近水体。四个性格各异的边界由一条环形步道串联，贯穿湖的四周。

为实现可持续发展的设计目标，设计师从多方面考虑了雨水和供水的问题。简单地说，所有不透水表面的雨水都通过各种形式进行收集，经过过滤和生态群落的净化以后储存起来，用于水景观供水。现场的雨水保持有利于缓解开发项目对外部水环境的不利影响，减少项目的净水消耗。同时，雨水的下渗还可以逐步改善场地中盐碱度过高的土壤。另外，现场雨水下渗避免了大量的现状雨水管道改造工作，为城市带来很大的经济效益。

天津文化中心是一个城市活动的核心，它把人们的活动与自然的环境相融合，并且激发人们去理解和体会人与自然的关系。中心湖周边丰富的景观环境不仅为现场的游客带来身心上的丰富体验，而且向整个城市散发出崭新的活力。

摄影：菲利普·内多姆莱 刘东
Photographer: Philipp Nedomlel, Liu Dong

Like the contrasting sounds of a symphonic crescendo and the soft smooth melody of a flutist's solo, the landscape experience in Tianjin's cultural centre captures the attention of its visitors. Following the result of the Urban Master plan design competition in approximate 3 years, an international design team composed of Landscape Architects, Urban Planners, and Architects has created a dynamic set of cultural spaces, indoor and outdoor which are connected and unified through the artful use of water as the core design element. A lake is at the heart of the park with an urban promenade at its edge that leads to the Opera, the park's focal point. An eco park, gardens, plazas, and terraces surround the lake and lead visitors to art, history, and to the existing natural history museum as well as to a new shopping mall and the youth centre. It is truly a rare outstanding example to show culture architecture group in the urban centre, meanwhile complete in limited period with high quality. The successes of Tianjin Culture Park takes Chinese government and project designers' responsibility and commitment to Tianjin citizens, therefore Atelier Dreiseitl design team is sincerely proud to participate this project and to be the witness of Chinese urbanization development.

This current master plan design phase focuses on the Culture Park itself. The site encompasses a 90-hectare area with a 10 hectare lake at its center. An Urban Promenade winds through the site for 700 meters with the most dramatic viewing point along the front of the opera house and lake at a length of 150 meters. The general design concept also includes intimately human scaled spaces allowing visitors in the park to enjoy the opportunity of reflection in the quiet and serenity of the gardens. The goal of the master plan was to create a cultural district of international esteem around the existing Natural History Museum. Yet, the emotive landscapes dedicated to promoting cultural pursuits also provide the simple pleasures of sport and recreation including skating areas, picnicking sites and a host of other opportunities for generations to interact. So, while the Culture Park is an international destination it is designed with the people of Tianjin in its heart.

While water is celebrated throughout the Tianjin Cultural Park in dynamic urban features, the lake with its natural character is the heart of the landscape experience. Like a painting in a museum, it reflects the human condition. Each edge of the framed lake relates to the space adjacent to it. The main promenade acts as an elegant urban edge where the water meets the city. The connection is further refined with sculpted wave patterns and reflected light. At the Tianjin Art Museum, a plinth resting just below the surface of the lake provides a dramatic exhibition area for outdoor installations. The Eco Park's shoreline is complex and natural in pattern with reeds, boardwalks, cleansing biotopes, and it is inviting interior water inlets. In front of the new Opera house, the generous steps as the city stage should provide the opportunity to descend to the lake surface and enjoy the water. All these interfaces are linked to a walking loop that circles the lakeshore.

The design approach to storm water and water consumption needs is multifaceted. Stormwater is collected from all the impervious surfaces on site, filtered and cleaned in cleansing biotopes and then stored for reuse in water features. Keeping the water onsite reduces the impacts to the natural water systems off site and the water consumption on site. Meanwhile, through the infiltration, the imbalance of the over salted water should be corrected in the area. The city is benefited by onsite infiltration in a financial way as well as the existing sewer system could not accommodate the additional runoff.

The Tianjin Culture Park is an urban destination, which engages the human connection to the natural world and inspires our research to understand and celebrate the human condition within it. The range of landscape experiences surrounding the central lake harmonizes together to feed the minds and souls of visitors, and also to spread a renewed creative synergy throughout the city.

全景鸟瞰图 Panoramic Aerial View

总体铺装布局
Overall Pavement Layout

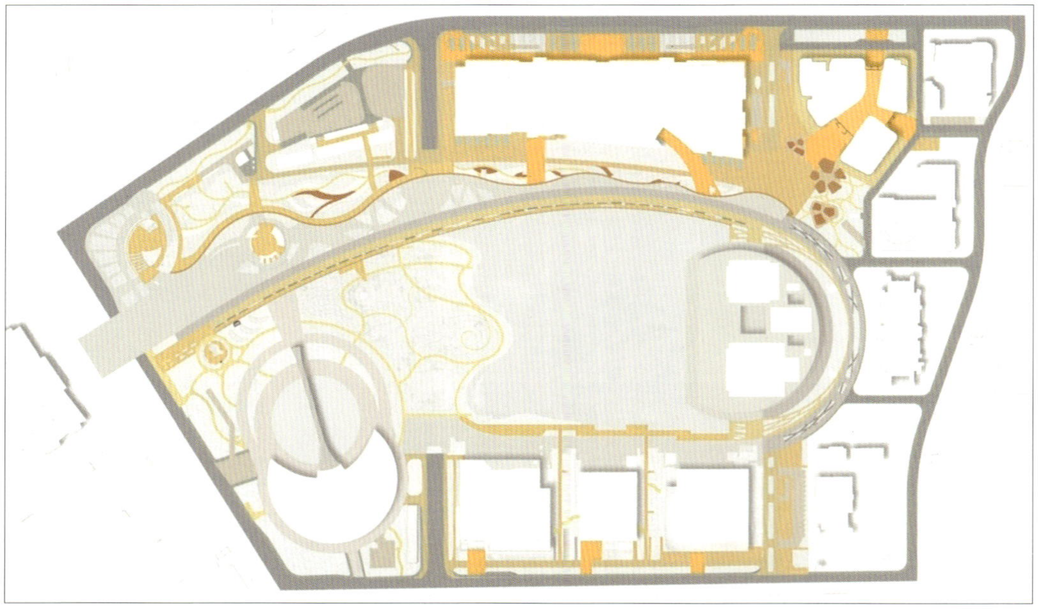

米黄色与灰色铺装设计清楚的定义出建筑周边区域与广场区域空间
The Beige and the Gray Pavement Define the Surrounding Area of Buildings and the Square Area

法桐步道与园路局部铺装
Plane-Tree Promenade and Partial Pavement of Garden Path

法桐步道
Platane Promenade

法桐步道
Platane Promenade

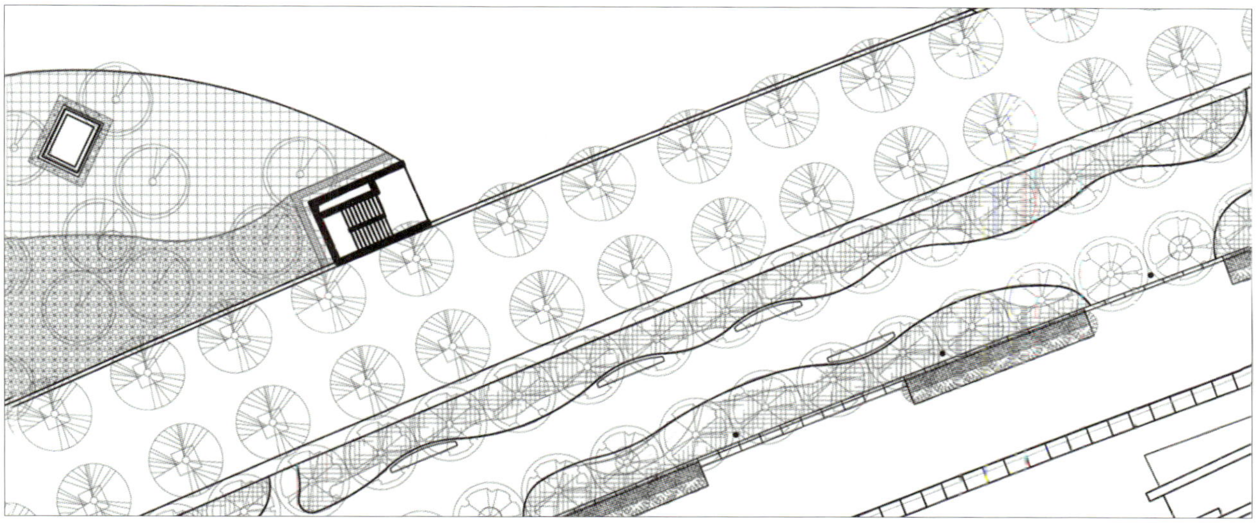

海棠步道
Malus Speetabilis Promenade

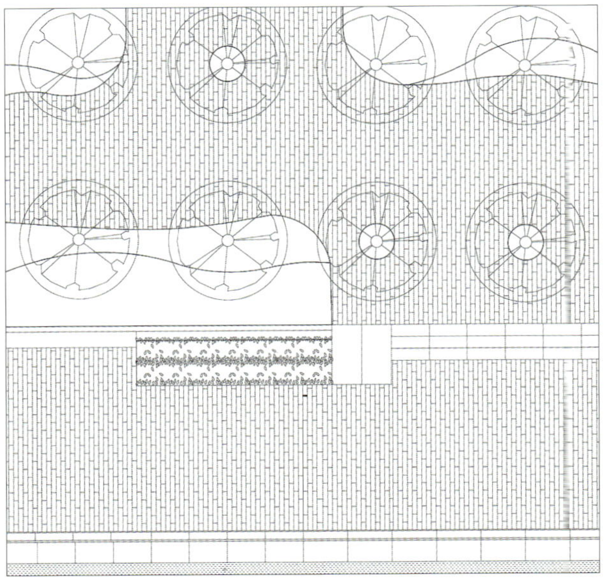

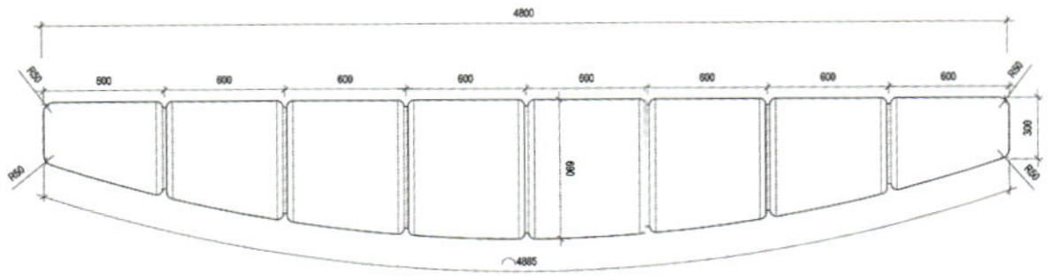

海棠步道
Malus Speetabilis Promenade

天津大剧院
Tianjin Grand Theatre
天津科技馆
Tianjin Science Museum
天津青少年活动中心
Tianjin Youth & Children Center
阳光乐园
Sunshine Plaza
市民广场
Citizen Plaza

滨水步道节点设计
Waterfront Promenade Design

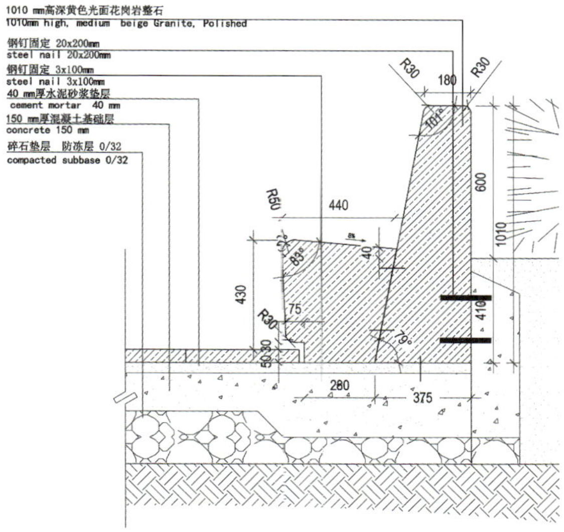

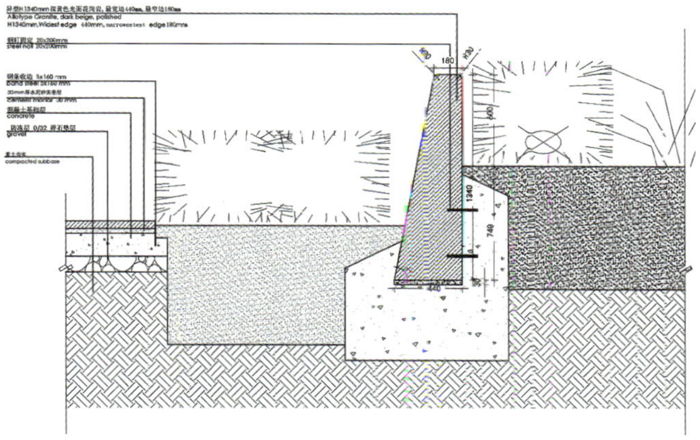

滨水步道
Waterfront Promenade

亲水驳岸节点设计

Waterfront Node Design

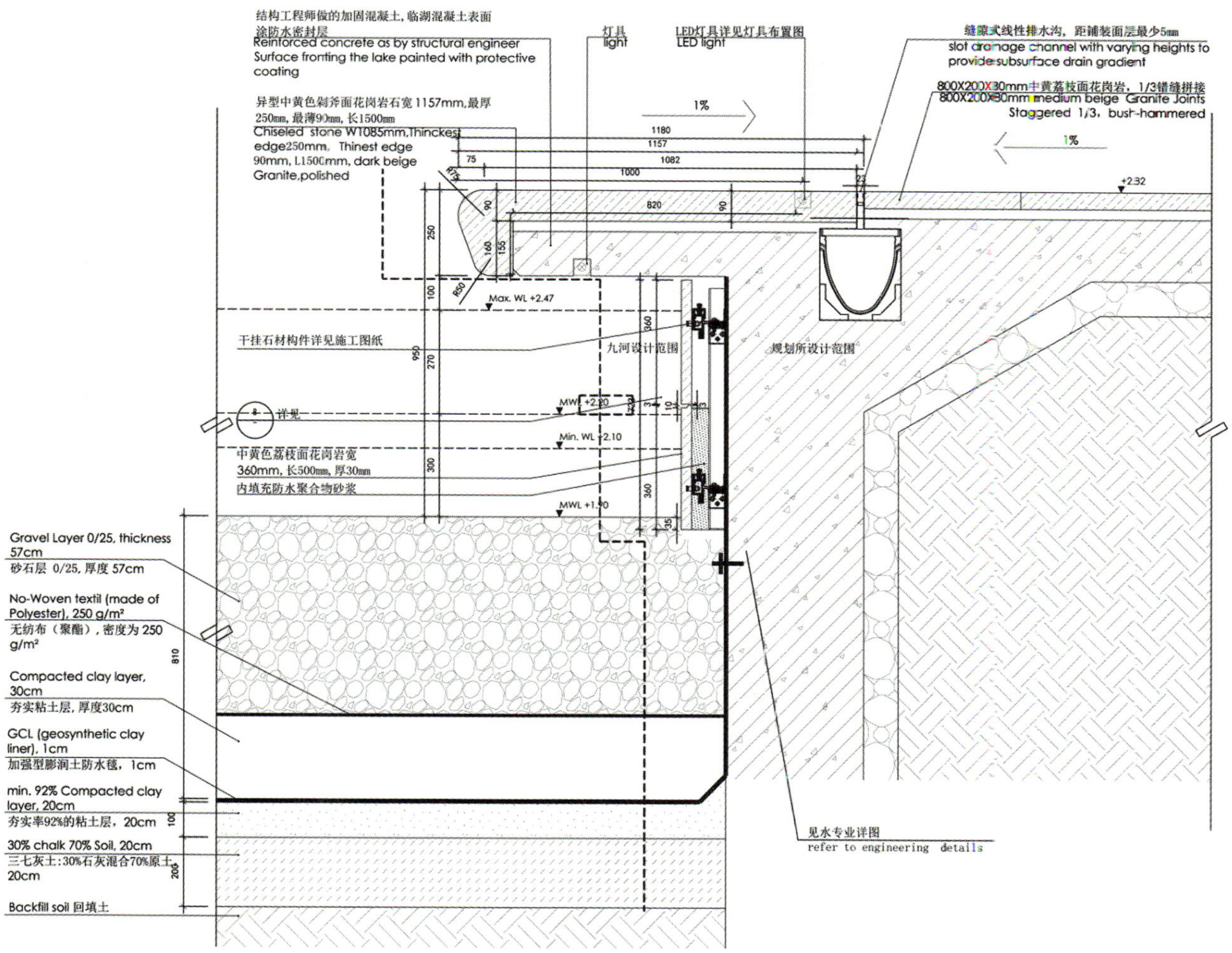

滨水步道夜景

Night View of Waterfront Promenade

角部下沉广场
Corner Sunken Plaza

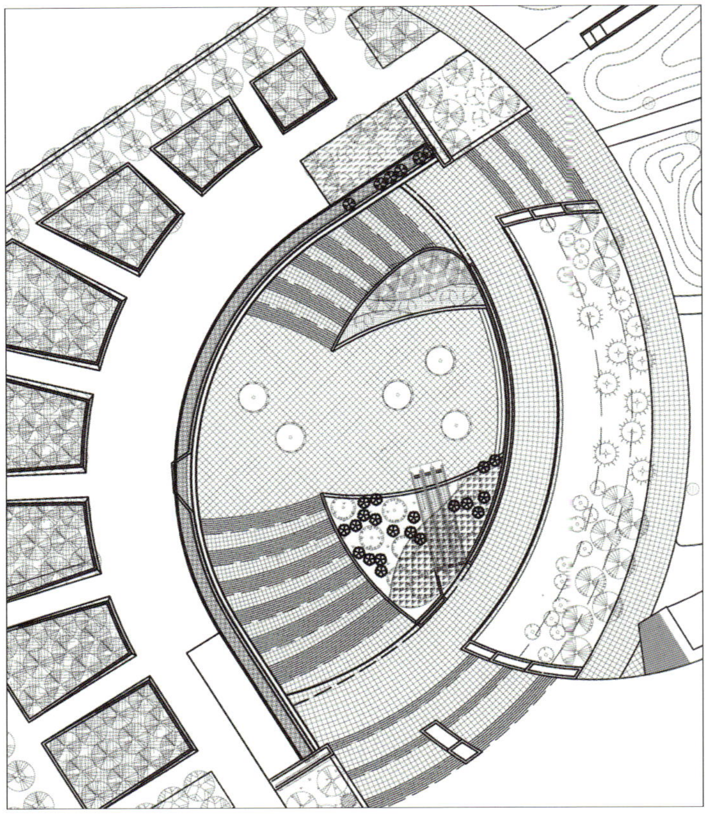

角部下沉广场
Corner Sunken Plaza

中央下沉广场
Central Sunken Plaza

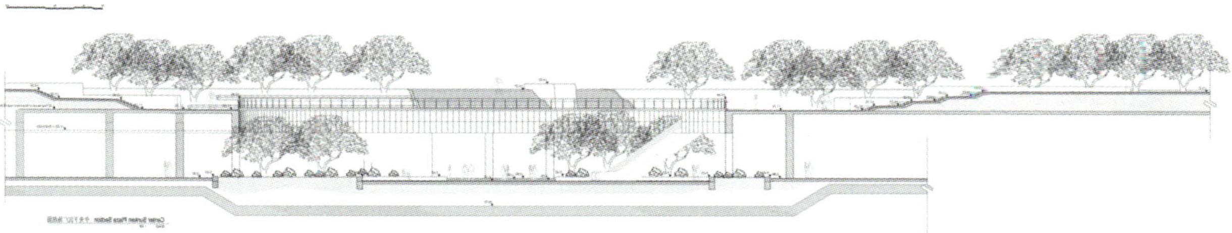

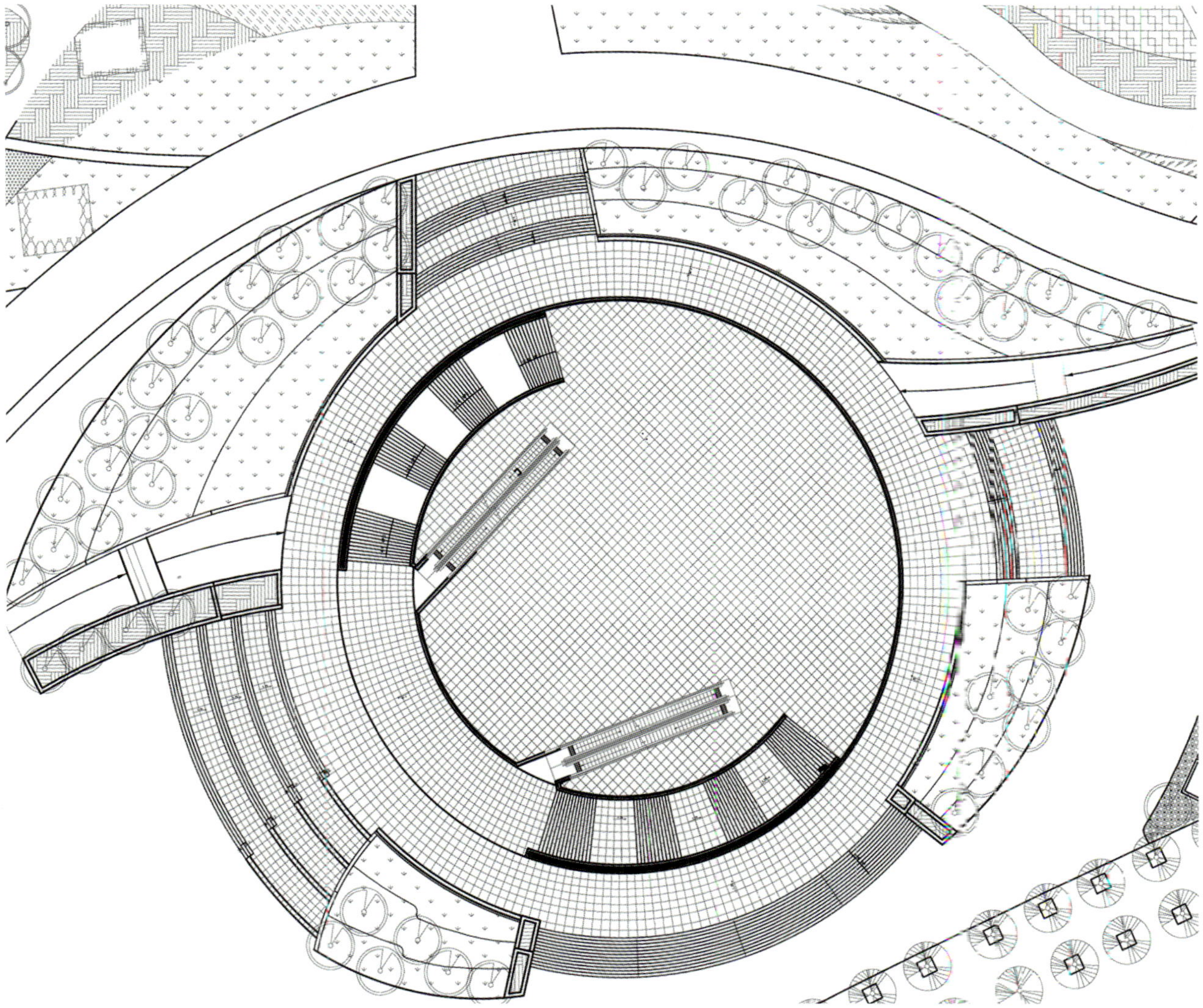

中央下沉广场
Central Sunken Plaza

花园下沉广场
Garden Sunken Plaza

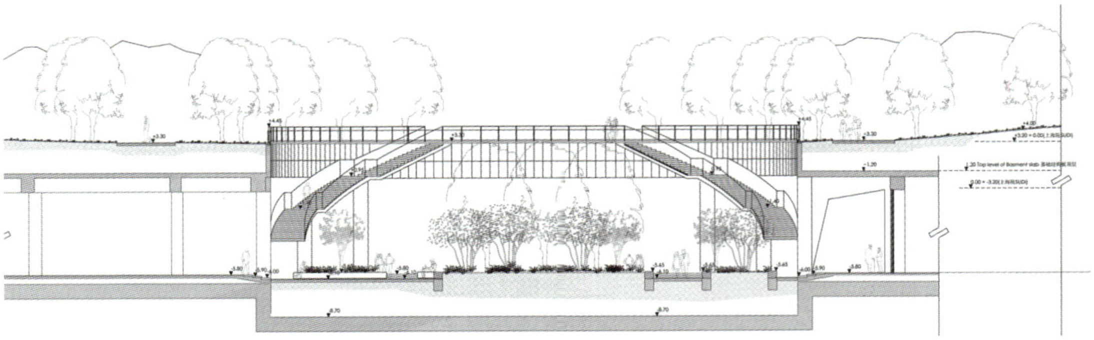

花园下沉广场
Garden Sunken Plaza

购物中心镜面水池
Mirrorlike Water in front of Galaxy Mall

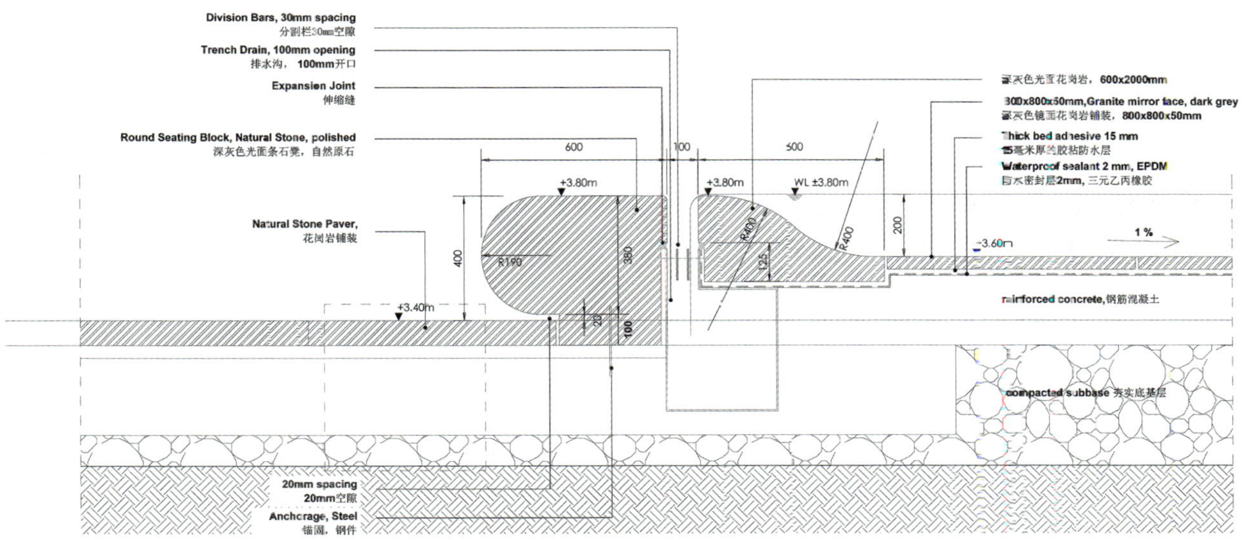

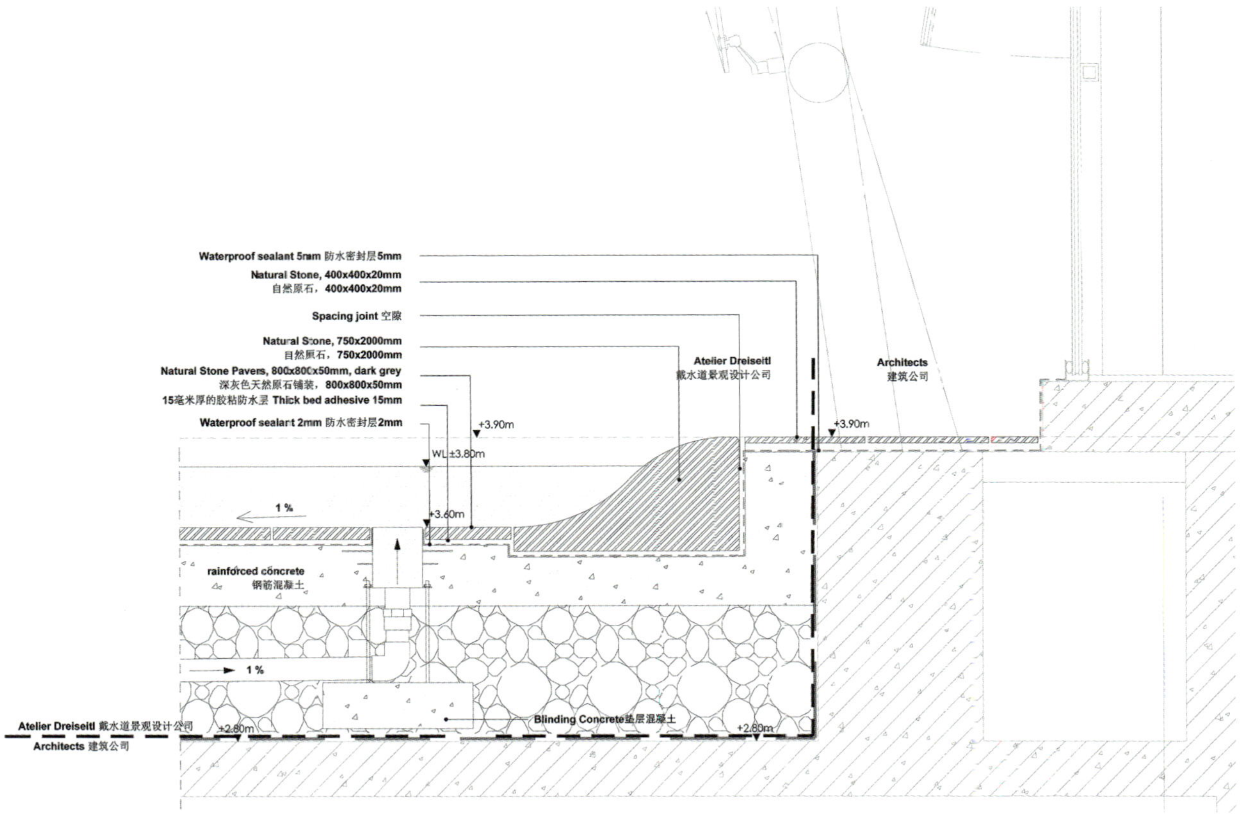

购物中心镜面水池
Mirrorlike Water in front of Galaxy Mall

文化中心周围法桐人行道
Platane Promenade around Cultural Center

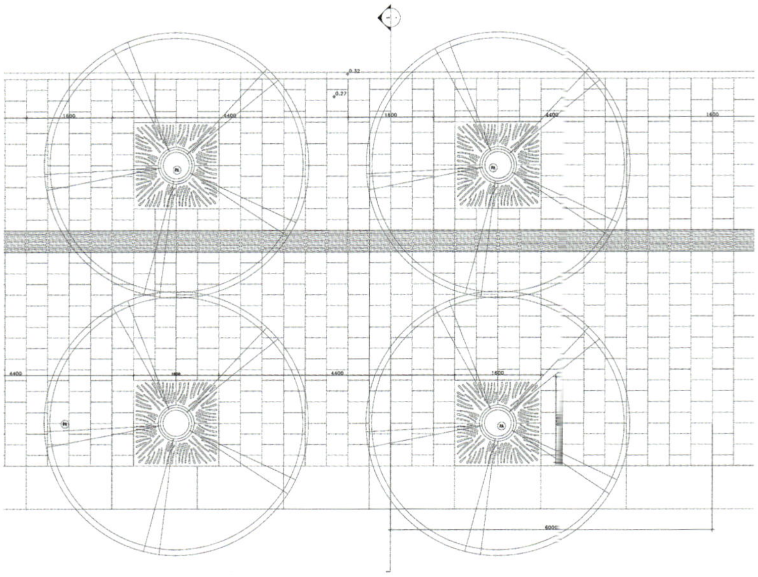

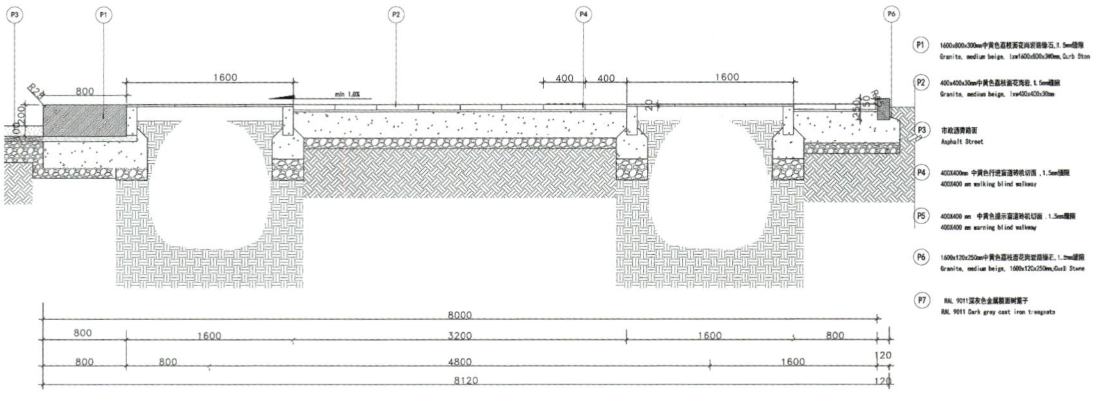

文化中心周边法桐人行道
Platane Promenace around Cultural Center

入口广场铺装基础及线性排水沟
Foundation of Pavement and the Drainage Ditch in Entrance Plaza

石材选样
Stone Selection

角部下沉广场
Sunken Plaza in Corner

角部下沉广场大台阶
Major Footstep of Sunken Plaza in Corner

步道广场 800X800 石材铺装
Paving the Walkway with Stone 800X800

海棠步道 200X800 石材铺装
Paving the Cherry-apple Tree Promenade with Stone 200X800

部分法桐步道铺装
Partial Pavement of the Plane-Tree Promenade

滨水步道节点
One Node of Waterside Promenade

海棠步道
Malus Speetabilis Promenade

步道花园内木平台
Wooden Platform on Walkway in Garden

海棠步道座椅
Chairs on Malus Speetabilis Promenade

通风口干挂石材
Dry Hanging Stone of Ventilation Opening

花园下沉广场
Sunken Garden

湖岸边异形石材
Dry Hanging Stone of Ventilation Opening

海棠景墙安装过程
Installation of Landscape Wall

花园步道 200X800 石材铺装
Paving the Walkway in Garden with Stone 200X800

购物中心镜面水池节点
Node of Mirror Pond Around the Galaxy Mall

管控中心木平台
Installation of Landscape Wall

生态水系统设计
Eco Water System Design

德国戴水道设计公司 Atelier Dreiseitl GmbH
天津市九河市政工程设计咨询有限公司
Tianjin Jiuhe Municipal Engineering Design Consultation Co., Ltd.

生态雨水系统
Eco Rainwater System

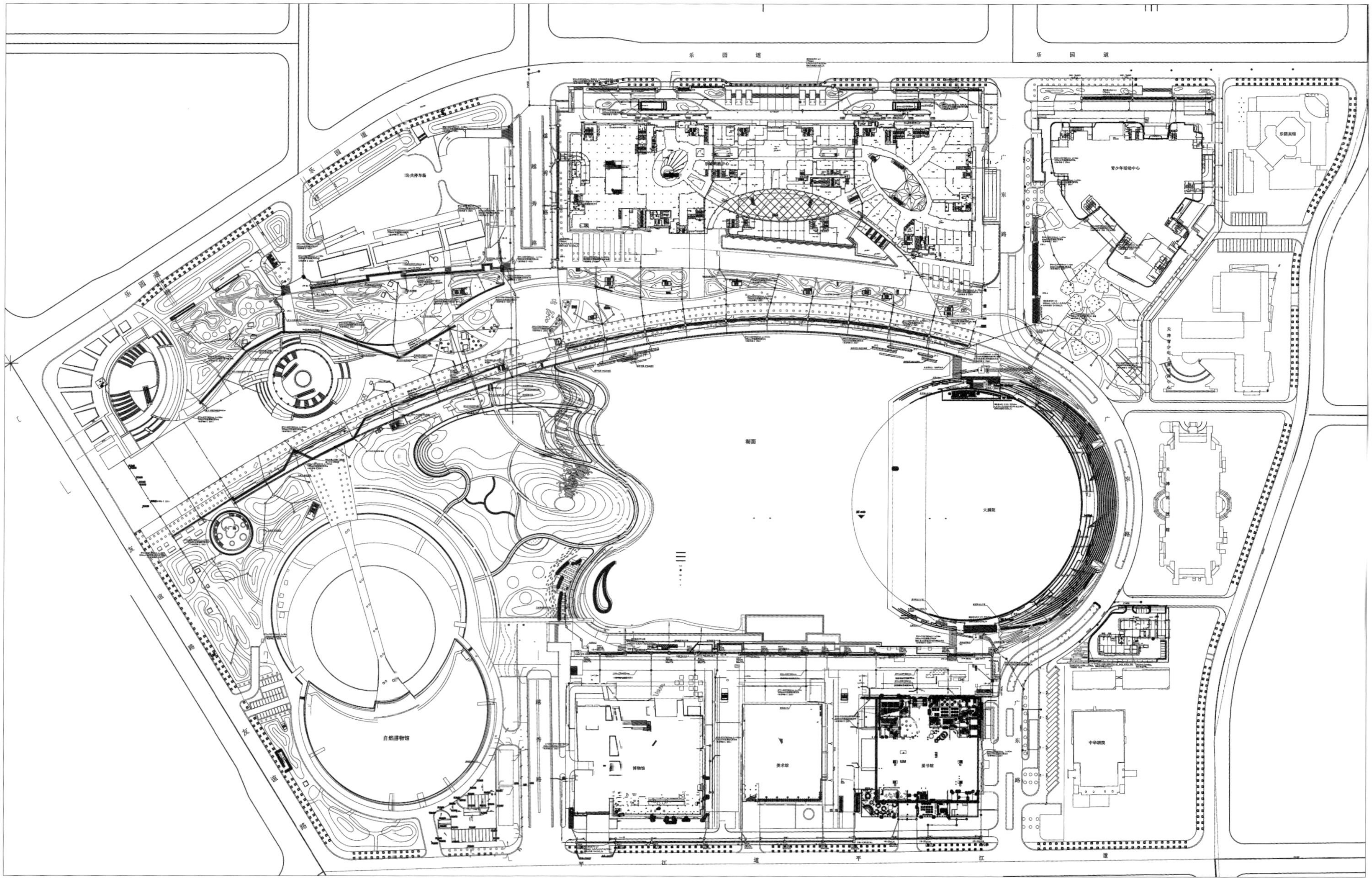

摄影：菲利普·内多姆莱尔
Photographer: Philipp Nedomlel

雨水收集系统

文化中心新建区大部分雨水可通过初期净化、滞留、储存后作为优质水源排入中心湖（内排区域），其余的雨水径流在排入市政管网之前，也会先经过初期净化和蓄水沟滞留，将对市政管网的压力减小到最低（外排区域）。超过设计标准的强暴雨径流可以通过蓄水沟的紧急溢流管道排入湖中或市政管网中。整个场地面积为90.09ha，共划分为22个集水/排水流域，设计重现期取P=3a，总滞留体积为7000m^3。相比于目前天津市普遍采取的0.5~1a的重现期，大大减轻了对市政管网造成的压力，并提高了防洪标准。整个场地外排雨水从24.8m^3/s消减至2.2m^3/s，紧急溢流到市政系统的流量为750 L/s（以10年一遇计）。雨水经净化后可用于补湖，年节省自来水费55万人民币。

Rainwater Collection System

Most of the stromwater is considered as a superior quality source of water which will be drained off into the centre lake after initial purification, retention and storage through massive decentralized channels and pipes. The rest of the stromwater should undergo a preliminary purification and should be detained in the ditch before being discharged into the municipal pipe network. That will decrease the pressure on the municipal network to a minimum. The strong stromwater flow exceeding the design standard can be discharged towards the lake or the municipal pipes via an emergency overflow pipe in the ditch.The total area is 90.09ha, divided into 22 Subcatchment areas. The design frequency is selected as P=3a, the total retention volume is 7000m^3 which largely relieves stress on the municipal drainage network and improves flood protection standard compared with frequency of 0.5~1a adopted by Tianjin at present. The volume of emergency overflow & deep drain into municipal systems is 750L/s. The stormwater used for feeding the central lake helps to save a cost of 550 thousand RMB per year. The peak outflow of the entire ground is reduced from 24.8m^3/s to 2.2m^3/s. The flow of emergency overflowing into the municipal system is 750 L/s(10 years return period). The rainwater is reused for refilling the lake after purification, it saves 550,000 RMB for one year.

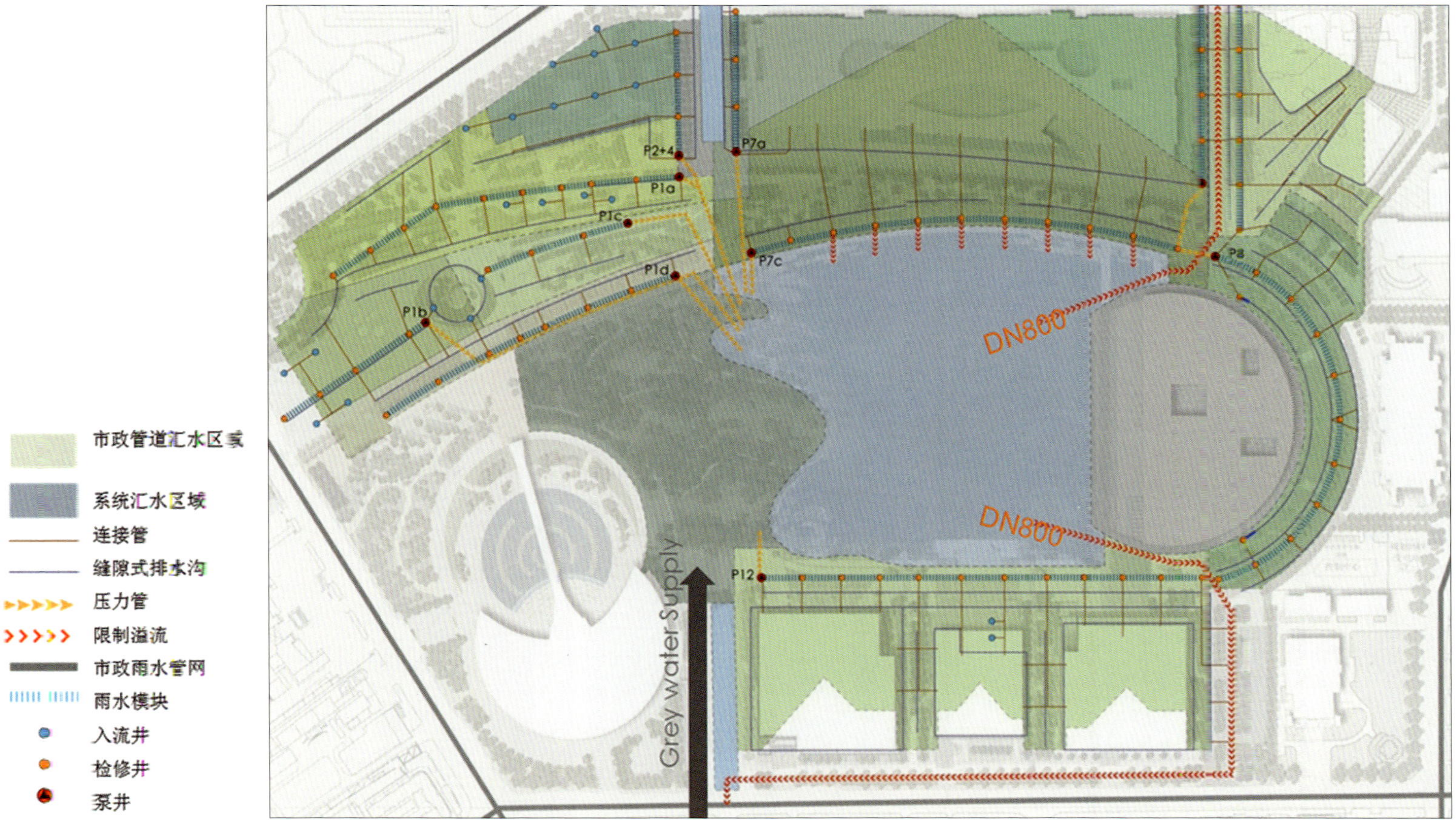

湖水循环与净化系统

中心湖的重要功能之一就是承接和分配无法外排的雨水，使其缓慢流入湖中。降雨中的高污染物粉尘含量将导致水中有机物浓度较高，这会加速湖体的富营养化水平，为浮游植物的生长提供有利条件，还致使藻类大量繁殖，最终使中心湖水体恶化，严重时甚至会出现湖水溢流并散发恶臭气味。因此，一个保障水质良好的循环与净化系统应运而生。一方面，湖体中的水不断地被抽出并进行处理；另一方面，雨水在入湖前也对其进行处理。这就确保了尽可能少的有机营养物有途径进入回流。因为浮游植物的生长和繁殖很大程度上取决于这些营养物质，尤其是磷起着至关重要的限制作用，根据研究结果，磷含量需要控制在0.03mg/L以下。然而雨水中的磷含量已经达到0.3mg/L，所以必须强调在入湖前对雨水进行处理。

循环净化目标是水质达到地表水环境质量标准III类水质。中心湖水体将通过撇渣器取水点、循环管线以及生态净化群落的流程以400m3/h的速度对湖水进行循环净化；另外处理房间的中水处理设备能够对补充的中水在进入生态净化群落和水体之前进行预处理，加药设备将对湖体进行生态处理方法以外的强化除磷处理。

生物净化群落是关键的设计内容，它是确保水质和净化功能的基础，包括水生生物的设计和过滤基层以及管道的设计

Lake Circulation and Purification System

The central lake will hold and dispense the incoming rainwater which cannot be drained away. The high dust content in the rainfall increase the organic input into the water . It could increase the lake's trophic level, and provide phytoplankton with excellent conditions for growing. This would result in rapid growth of algae, and deteriorate the appearance of the central system, which, in the worst case scenario, could lead to the water overflowing and creating a strong smell. For this reason, a circulation and purification system guaranteeing the water quality has been devised. On one hand, the water will continually be drawn from the lake and treated, and on the other hand, the incoming rainwater will be treated before entering the lake. This will ensure that as few nutrients as possible could find their way into the circuit. The phytoplankton heavily depends on these nutrients to grow and multiply; phosphor particularly plays the limiting role in this case. A phosphor concentration below 0.03 mg P/l is thus aimed for. However, the rain already contains 0.3 mg/l, which highlights the importance of treating it before it enters the lake.

The aim of water recycling and purifying is to reach class III in national quality standard for surface water. The centre lake will circulate and purify the lake water with a constant rate of 400m3/h through skimmers, pipe loop and the processing of cleansing biotope. Besides, the reclaimed water treatment equipment in the treating room will pretreat the reclaimed water before flowing into the cleansing biotopes and the lake. Trophical control equipments will activate extra P-removal treatments besides the ecological purifying steps.

The cleansing biotope is the key design component, including the design of the aquatic plants, filter substrates and pipelines

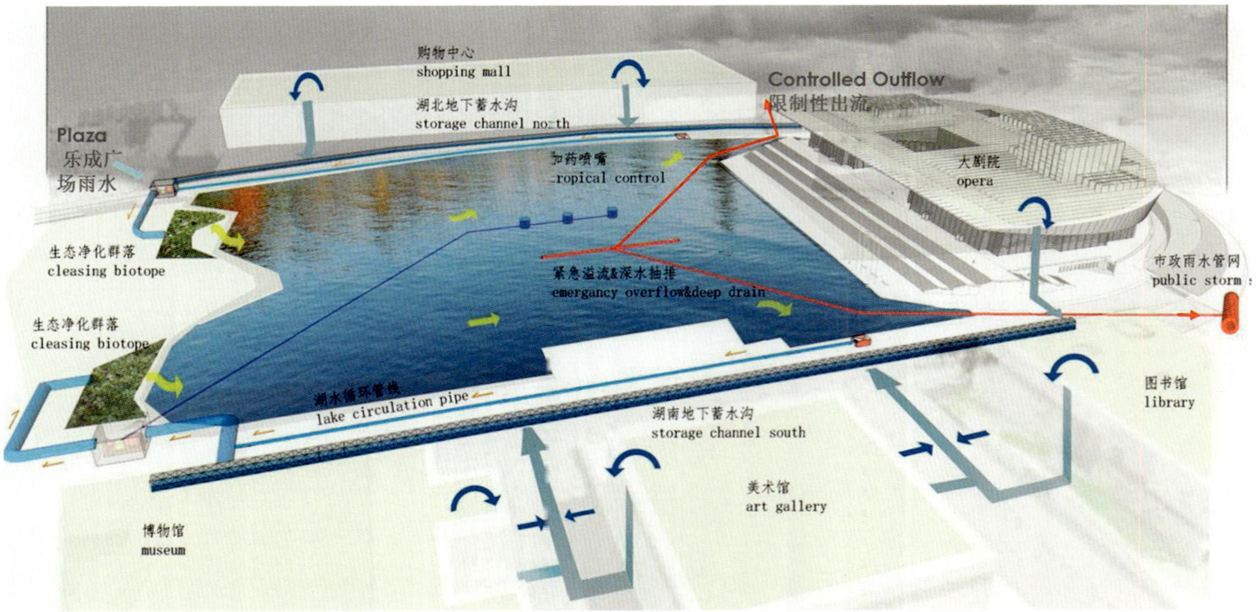

湖体设计

湖体形态设计由计算机模型生成并优化。模型考虑了水量平衡、水力动态和富营养控制各个方面的参数，例如中水补给量、收集雨水量、磷含量逐月负荷、浮游生物含量、沉淀速度等，建立了一个具有自然自净能力的三维形体，并与驳岸的设计相结合完成了最终三维形态和等高线。

湖体正常水位2.2m，最高水位2.5m，最低水位2.1m。正常水位距南北侧护岸顶高度0.5m左右，高差适宜。湖体完成面非常平缓，方便防渗材料的铺设。 湖底部约1.6公顷的中心区域水深3米，为湖的最深处。

湖体需采用粉末状天然钠基膨润土防水毯（以下简称GCL）进行防渗。

湖底防渗标准构造从上至下为：

① 0~25mm粒径的沙砾覆盖层60cm

② 密实度92%的现场粉质粘土保护层50cm

③ 粉末状天然钠基膨润土防水毯

④ 密实率92%的现场粉质粘土下垫层30cm。

Lake Design

Morphology design of the lake is established and optimized by computer modeling based on the parameter of water volume balance, hydraulic cynamic calculation and eutrophication control. The target is to produce a three-dimensional graphic with natural self-purifying capacity which is also well-merged with the form of embankment. Gentle sloping of the lake body facilitated the elaborate sealing layer.

The normal water level in the lake is 2.2m with highest 2.5m and lowest 2.1m. Deepest point at the bottom of the lake is around 3 meters from lake surface.

The sealing layer adopted GCL materials. From top down:

① 0~25mm Gravel and sand top layer 60cm

② Clay layer 50cm, 92% compacted

③ Natural and powdered sodium bentonite GCL layer 1mm

④ Clay base layer 30cm, 92% compacted

雨水收集系统
Rainwater Collection System

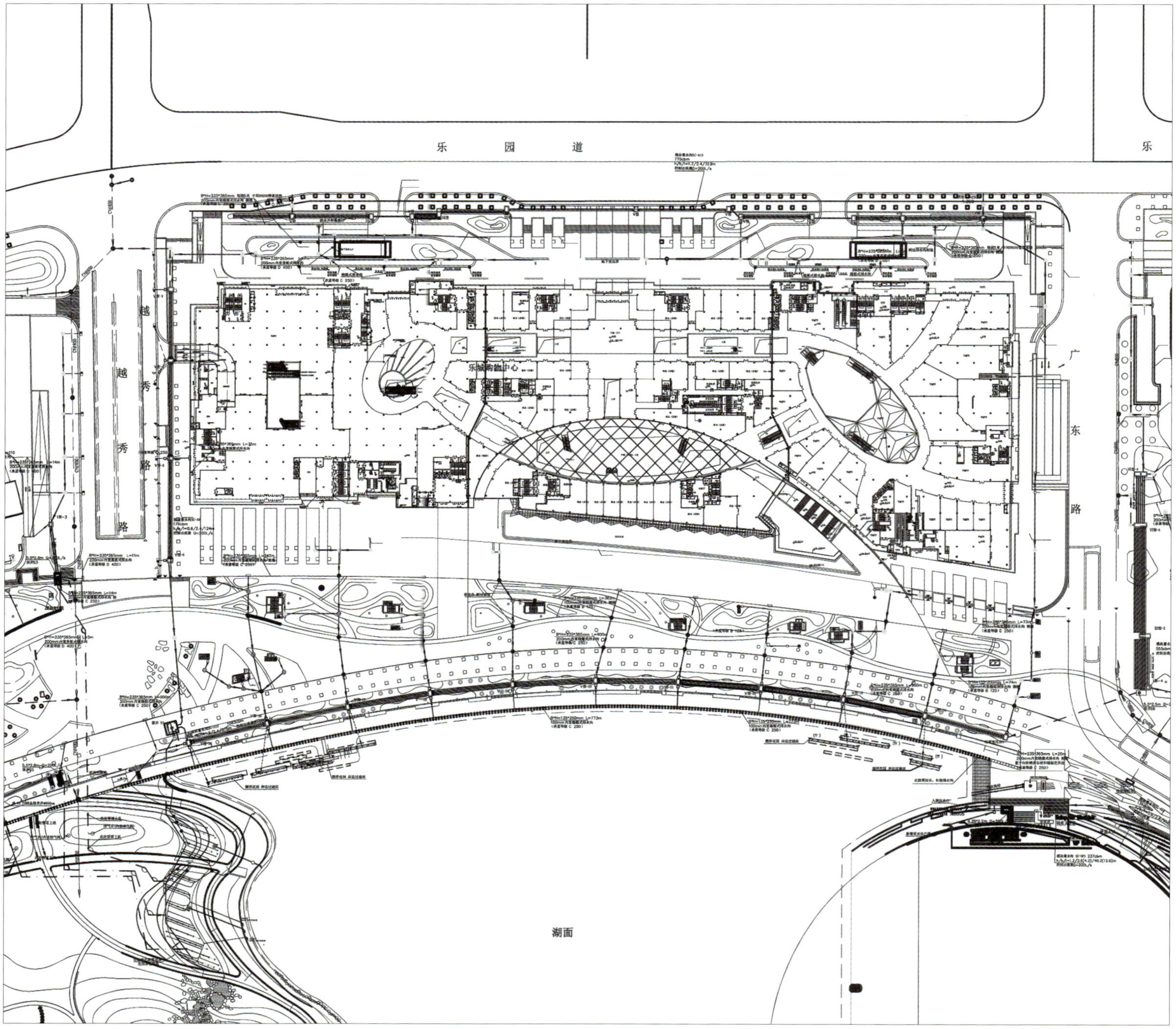

湖水
Lake Water

雨水模块
Rainwater Module

雨水模块连接检查井
Rainwater Module Linking an Inspection Shaft

沉淀管
Sedimentation Pipe

入流井
Seepage Well

沉淀井
Sedimentation Pipe

线性排水沟
Linear Drainage Ditch

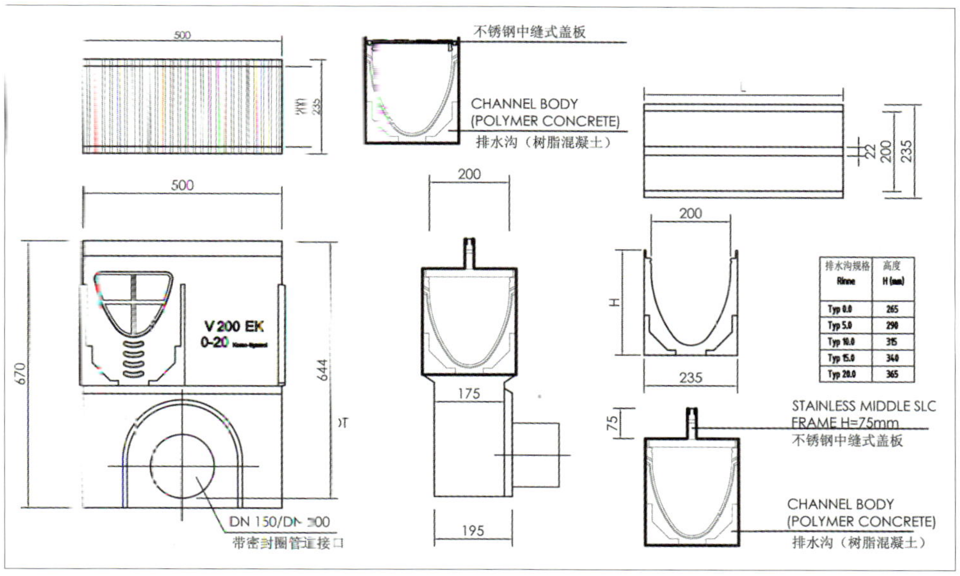

线性排水沟
Linear Drainage Ditch

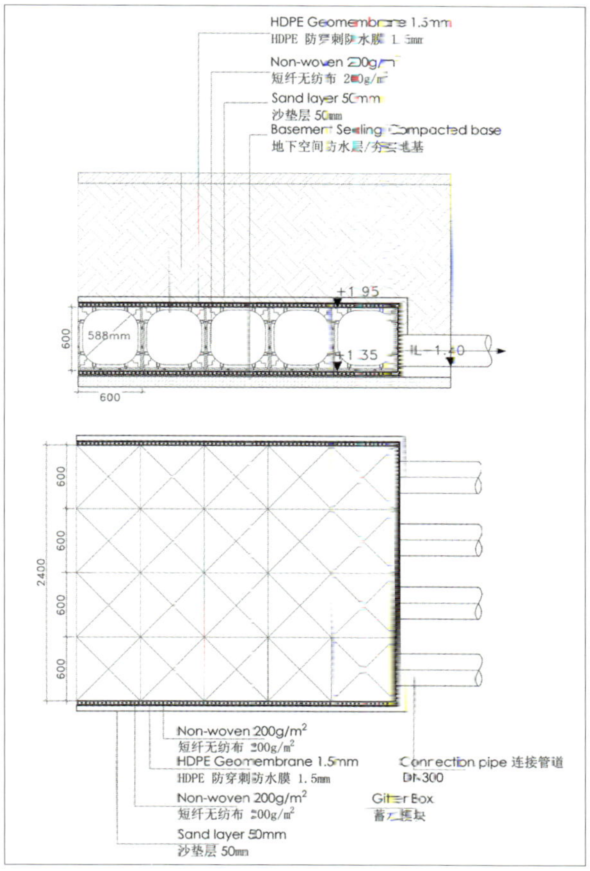

蓄水模块
Storage Module

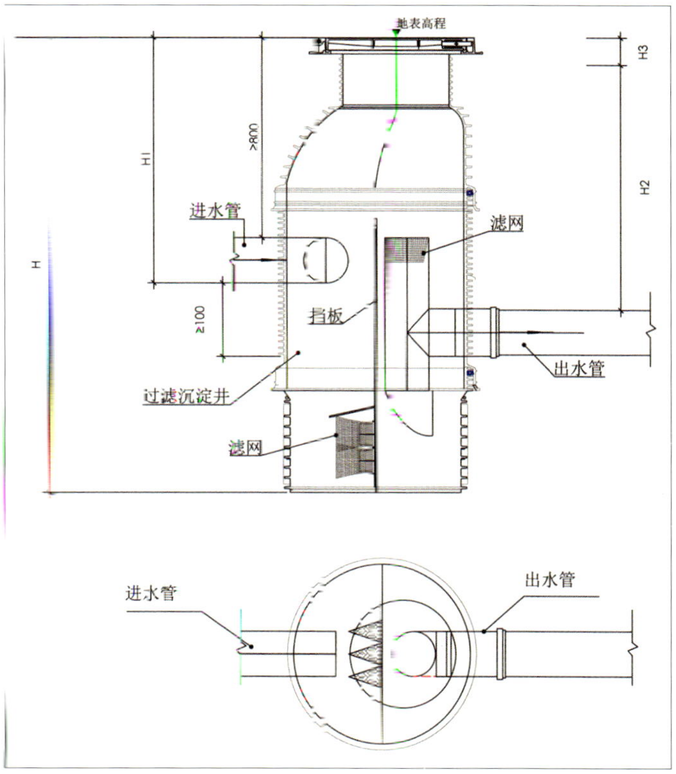

蓄水模块检查井
Inspection Shaft of Storage Module

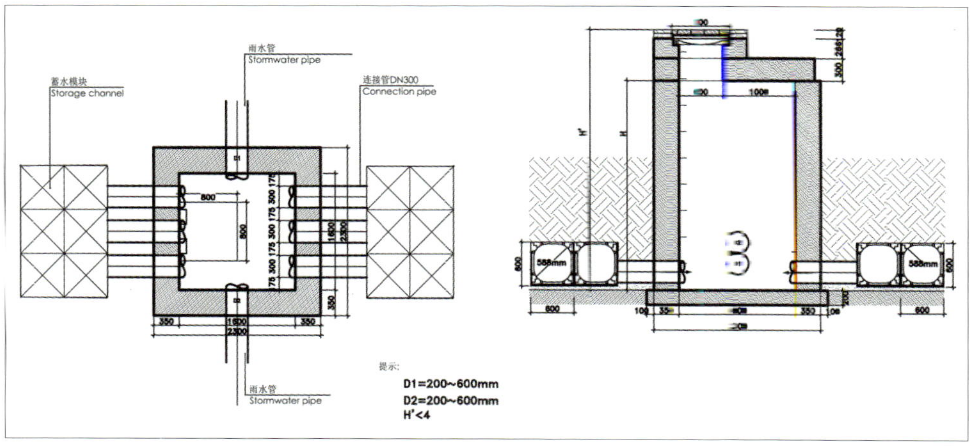

沉淀井
Sedimentation Well

湖水循环与净化系统
Circulation and Purification of Lake Water

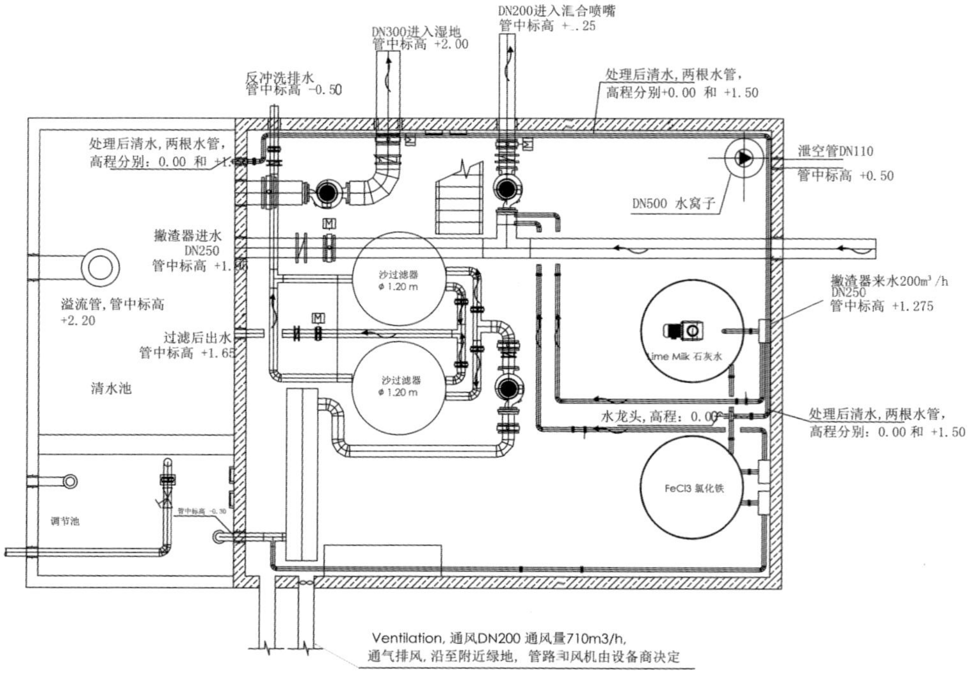

中水处理间
Reclaimed Water Treatment Room

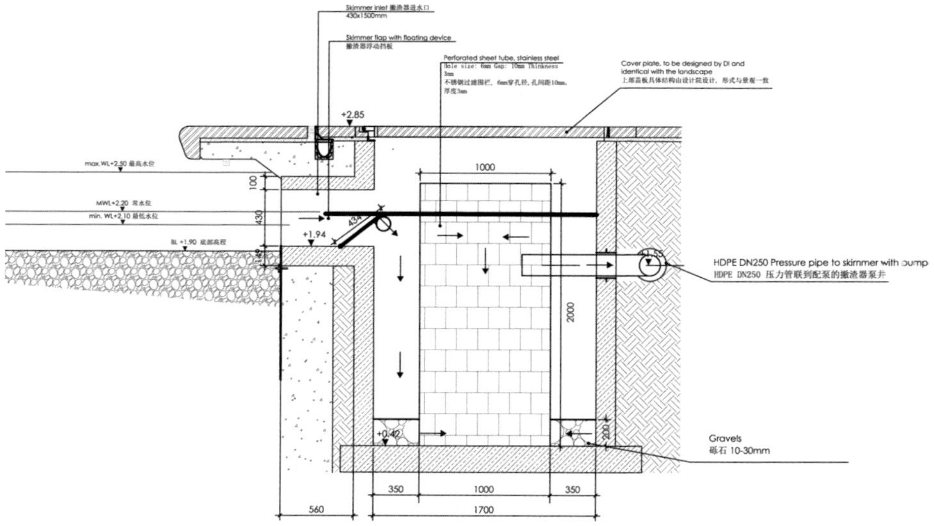

撇渣器
Slag Skimmer

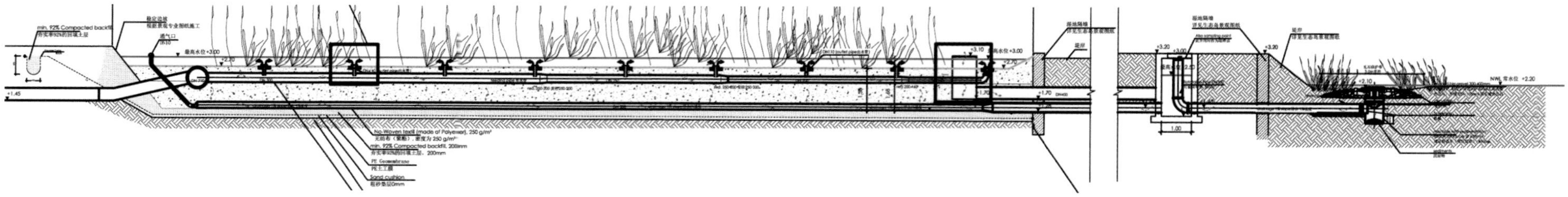

南部湿地
The Wet Land in South

循环与进化
Circulation and Purification

湿地景观
Wetland View

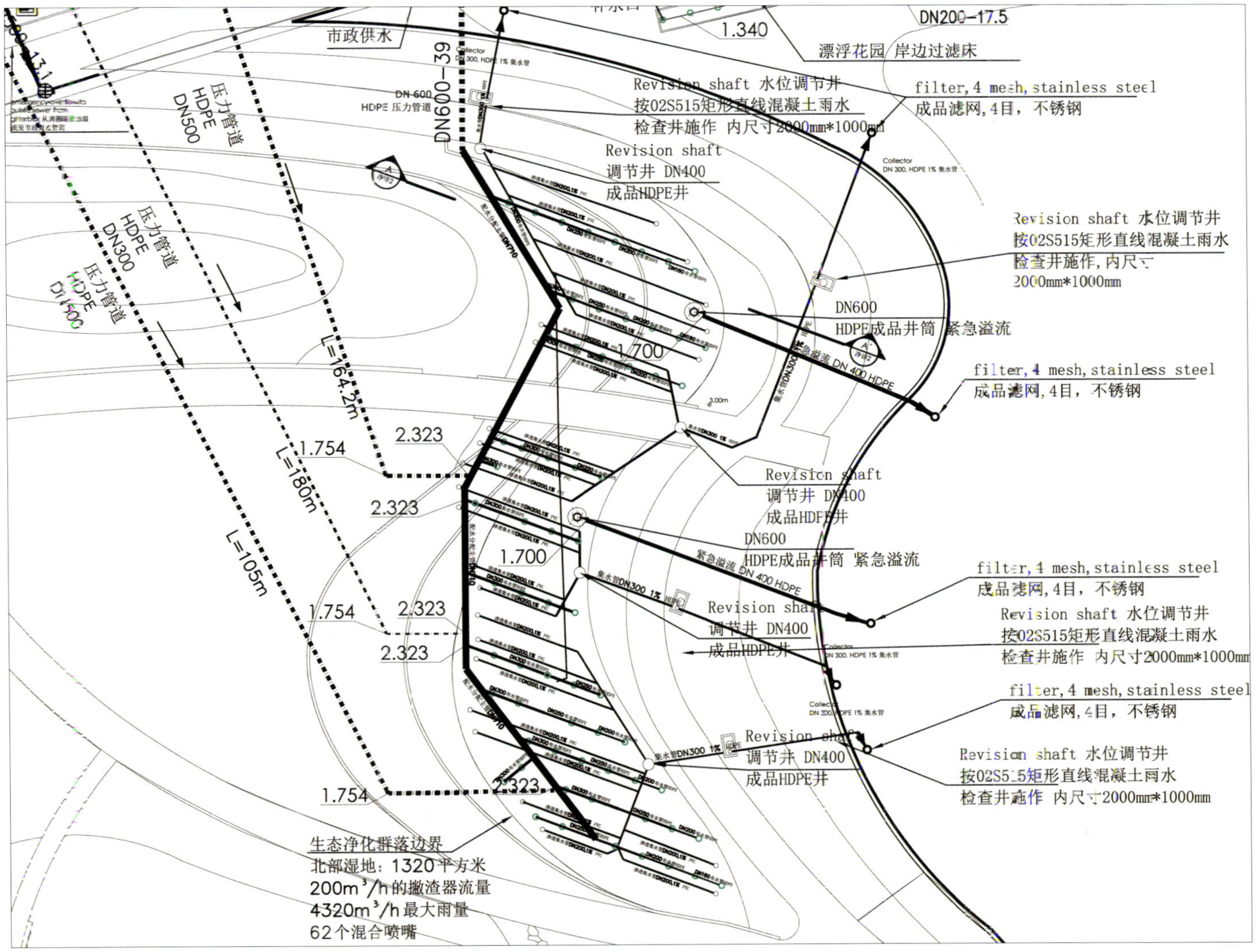

北部湿地

Northern Wetland

湖体设计
Lake Design

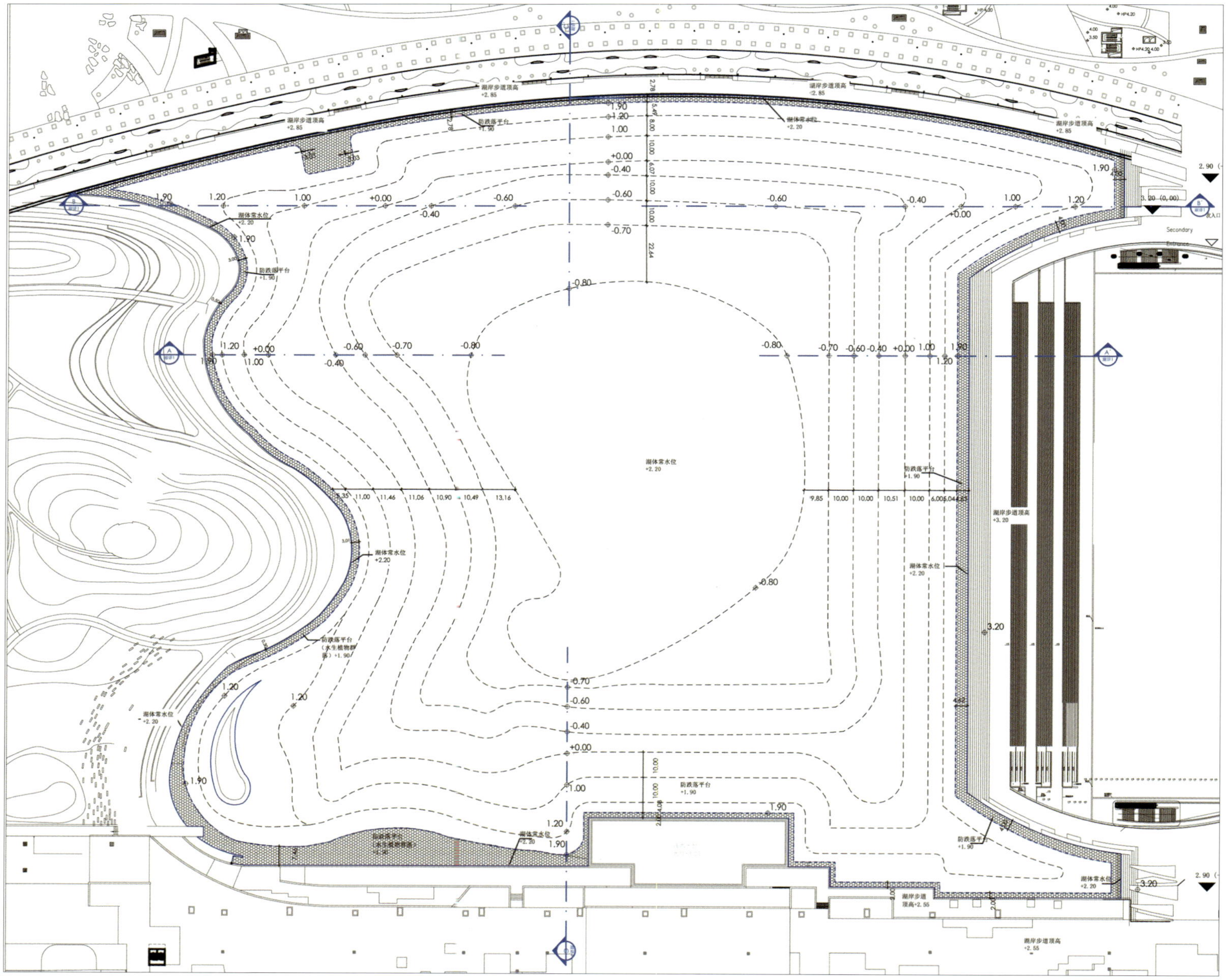

湖体高程设计
Lake Elevation Design

南岸设计
South Waterside Design

湖体设计 Lake Design

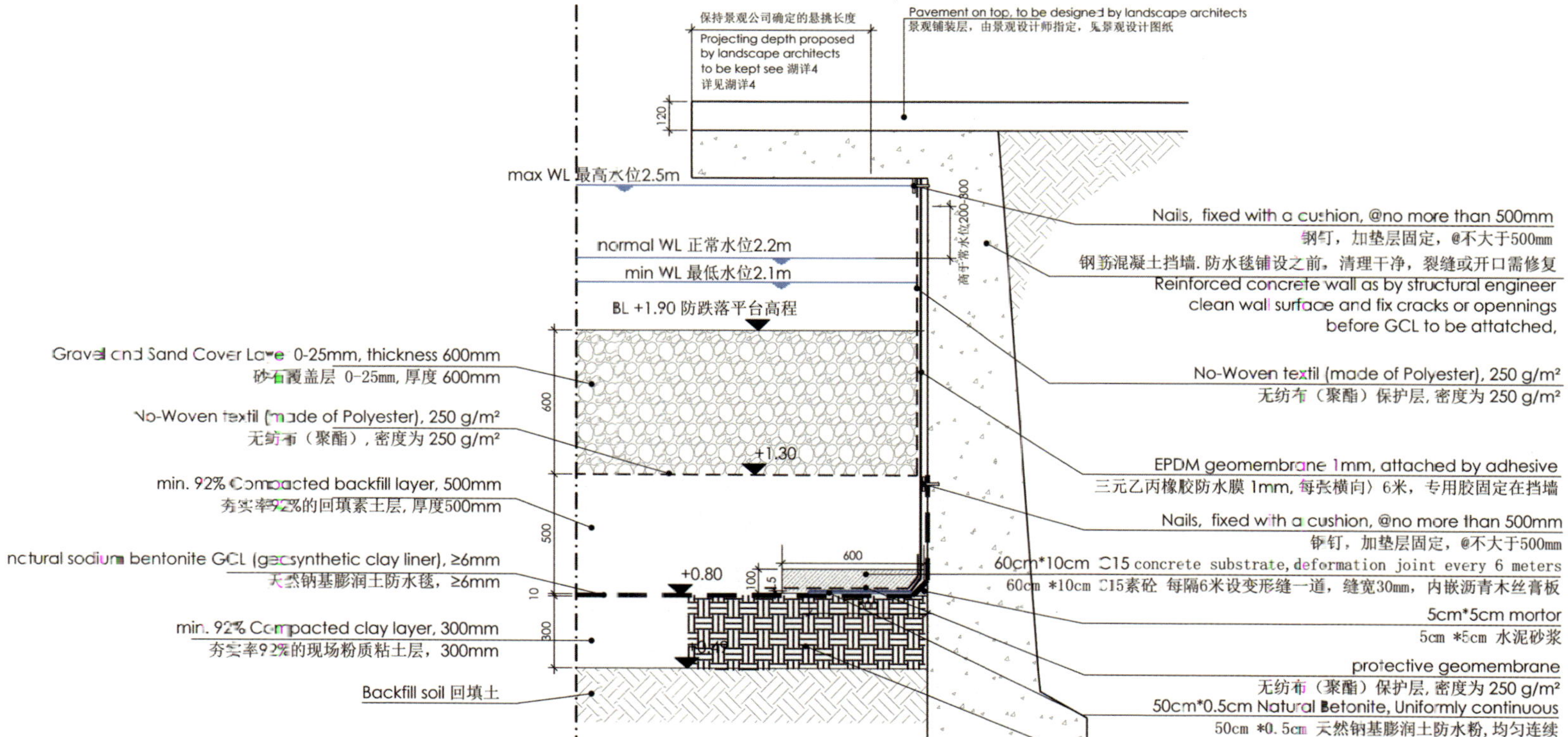

防渗设计
Seepage Control Design

迎宾塔设计 Yingbin Tower Design

天津市建筑设计院 Tianjin Architecture Design Institute
川口卫建筑结构设计事务所 KAWAGUCHI & ENGINEERS + EASTERN design office
天津大学建筑工程学院钢结构研究所
The Steel Structure Research Center of Civil Engineering School of Tianjin University

摄影：孔翠妍、曲可鑫
Photographer: Kong Cuiyan, Qu Kexin

建筑设计

天津市，位于北京东南方115公里处。在天津市内步行很容易迷失方向。历史上的天津和现在的天津有着相似的复杂性。历史上，天津是海外势力登陆中国的地方，欧美文化和中国文化曾在这里交汇，同时，作为首都的咽喉所在，这座城市也在发生着各种各样的变化。

在复杂的街道中，市民和政府能够会面。在现代化的高级公寓、大学和政府建筑物的旁边，砖造的居民区并列两旁。同时街道上宽裕的地方有人开设店铺。这座城市生动的表情，是与北京、上海这些大都市不同的地方。

Architectural Design

Tianjin is located 115 kilometers southeast to Beijing. Walking in Tianjin can easily get lost. The history and the present of Tianjin have the similar complexity. In the ancient time, Tianjin was the place in which overseas forces landing China, Europe and the United States' culture and Chinese culture interchanged in here with local one, meanwhile as the capital's throat, all kinds of variations are taking place.

In the complex streets, the public and the government can meet. Besides the Modern service apartment, University and government buildings, brick residential buildings are on the both sides. At the same time someone open stores in the well-off street. The city's vivid expressions differ form other metrololitans such as Beijing, Shanghai.

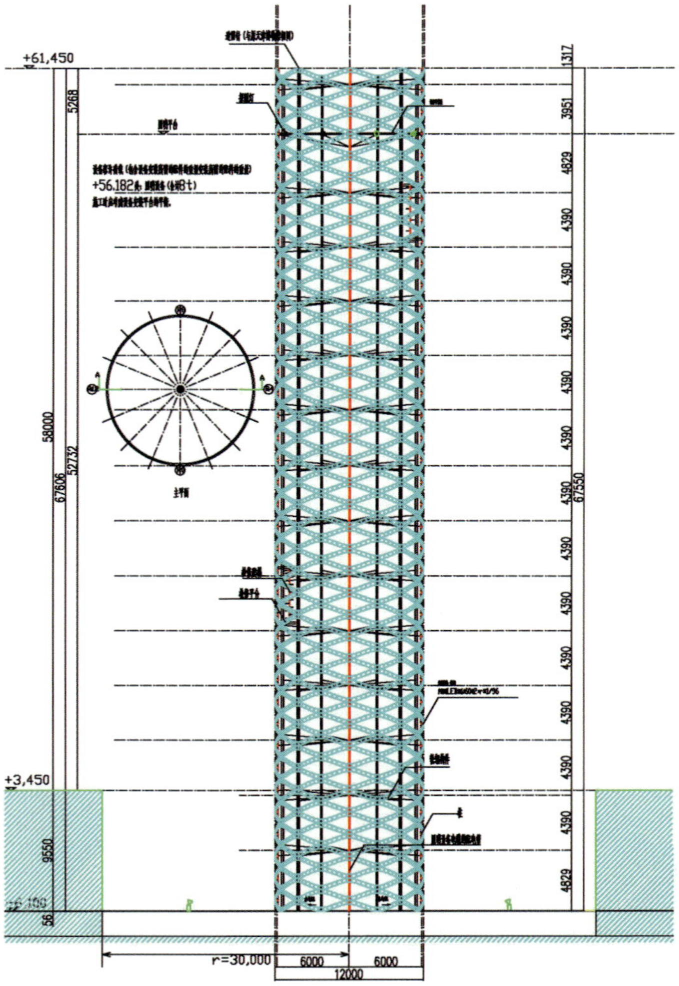

迎宾塔景观
Yingbin Tower View

基本情况

这座塔矗立在中国天津文化中心广场。广场位于天津市大礼堂前方。这里对于天津市而言是相当于天安门广场一样的地方。这座塔就位于张贴石面的巨大广场上，称为“迎宾塔”。

塔直径为12米，除了日常维修以外，不允许攀爬。这座塔从广场以下6米的地面开始建立，自最低的地面起，塔高64米。

埋入铸钢构件的LED照明灯可以实现各样的夜景效果。年末的倒计时（显示变化的数字）是这座塔的重要作用之一。

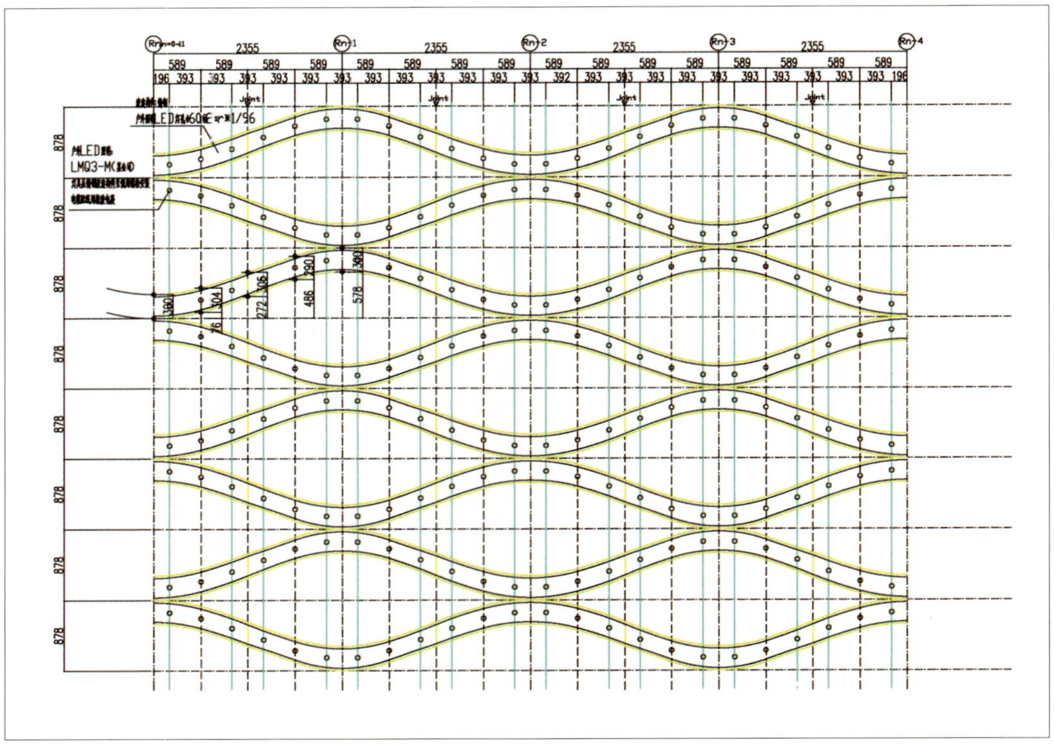

Basic Information

The tower stands in the Tianjin culture center, which is located in front of Tianjin Auditorium. The auditorium in Tianjin is equivalent to Tiananmen square, and the tower is located in the great square coverd by tiles, known as the " Yingbin tower ".

The tower is 12 meters in diameter, except the daily repair, it is not allowed to climb.

The tower start from the square 6 meters below ground and the height is 64m since the lowest floor.

LED lights buried in steel cast can achieve a variety of nightscapes. At the end of the year, The countdown (showing changes in digital) is one of the most important functions of the tower.

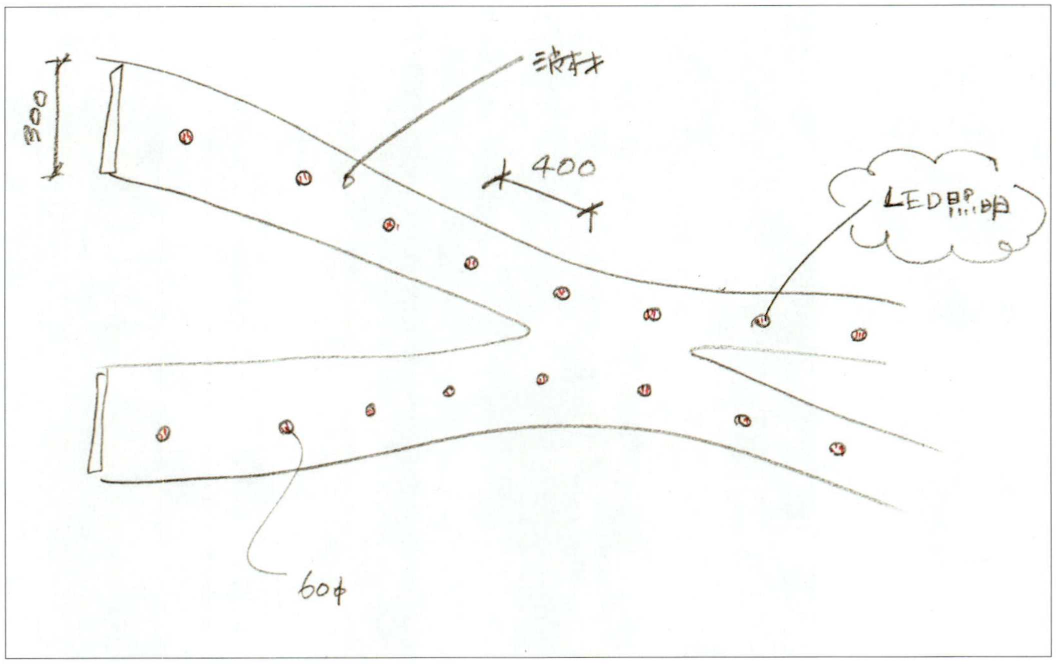

LED 照明设计
LED Light Design

迎宾塔夜景
Night View of Yingbin Tower

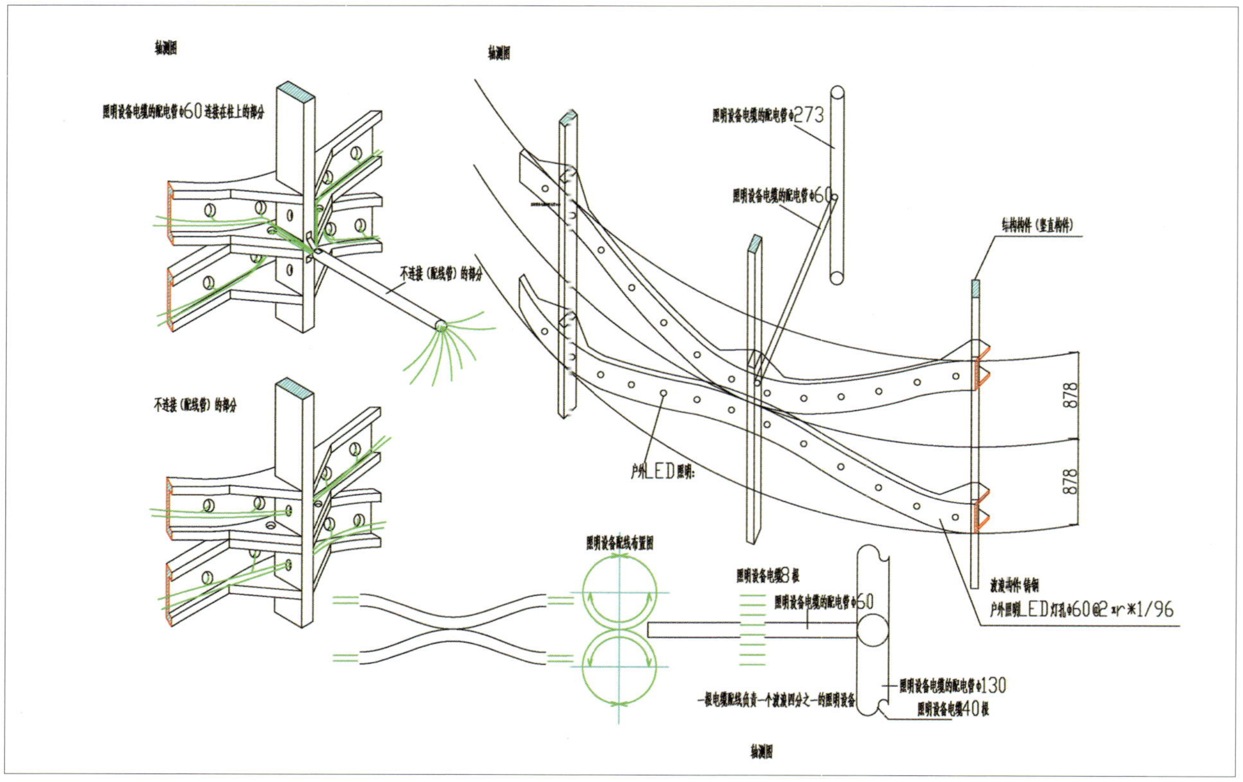

铸钢构件
Cast Steel Member

设计和结构

这座塔是世界唯一一座全铸钢建造的塔。

从真正的设计2011年12月31日开始到塔的完成2012年4月23日，只用了4个月的时间。

为了满足短期内完成施工的要求，选用了铸钢材料。

该塔共有1136个*形构件。假定4个*形构成一个单元，则一个浇筑模具需要浇筑284个构件，铸造模具的成本几乎不影响整体的造价。

* 沿着圆形弯曲，也就是说构件是弯曲的。铸钢可以很好地实现这个形状，像起伏的波浪一样的圆环共有71层，层层堆叠。

Design and Structure

All the material used to build this tower is cast-steel. This is the only tower in the world that cast iron is exclusively used for the construction. This is the number one tower in the world.
It was on December 31, 2011 when we started to make first drawings. The completion of the tower was April 23, 2012. From the start of planning, only four months were given until completion. Cast-steel is used to meet such a limited construction period.

1,136 pieces of wave* shape pattern members are used for this tower. In case 4 pieces of wave* shaped cast-steel members make one member, 284 members can be made with only one tooling, so the tooling cost does not affect too much to the total building cost.

Wave* shape iron steel is a piece of a circle, and it has no straight part. It is twisted; this can be realized with cast steel. 71 of wavy levels are piled up to build a big cylinder shaped tower.

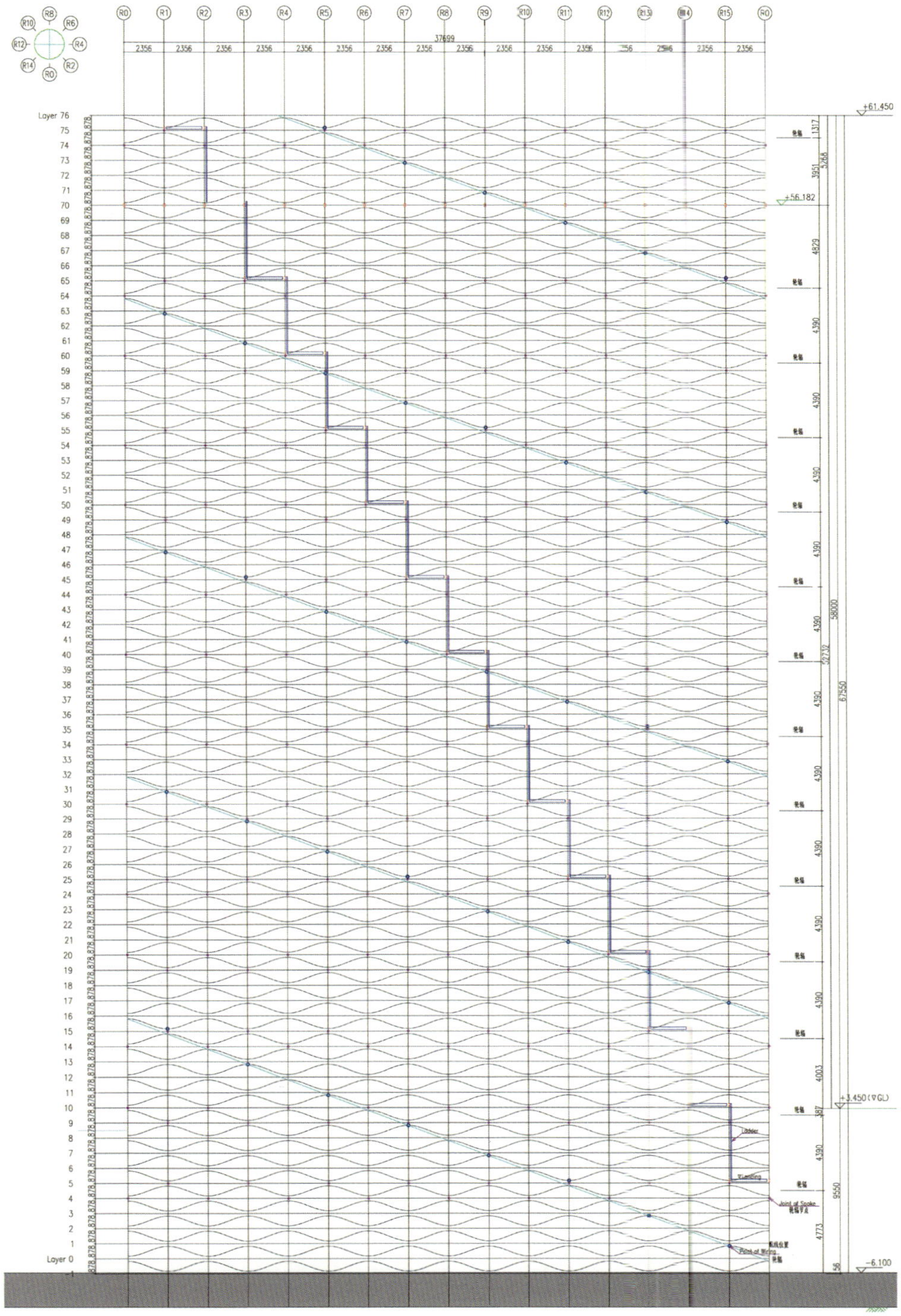

结构展开图
Structural Expansion Drawding

迎宾塔
Yingbin Tower

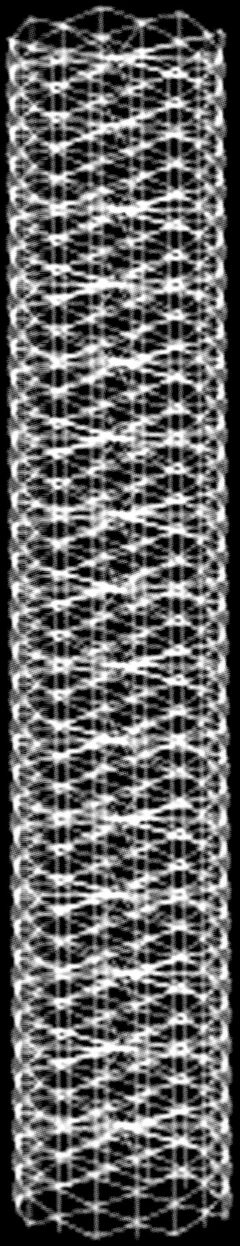

结构整体模型
Structural Model

柱布置
Columnar Layout

波浪布置
Wavy Layout

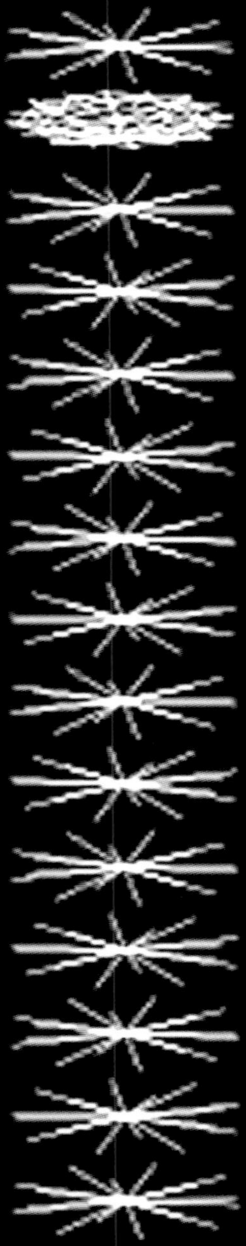

轮辐构件及平台布置
Spoke Structure and Platform Layout

结构设计

针对迎宾塔工程建筑造型上的特点，采用铸钢构件可以实现其波浪圆滑过渡，且铸钢节点在工厂预制，现场焊接量小，同时采用铸钢节点可以满足该工程短期完成施工的要求，大大节省人力、物力和建造时间。

铸钢件在工厂预制，并进行机械加工和热镀锌处理。经处理后的铸钢表面平整光滑，使迎宾塔更像是一座艺术品。

塔体构件采用铸钢节点材质均为德标G20Mn5。本工程共有624个基本节点，每个节点重约0.7吨。

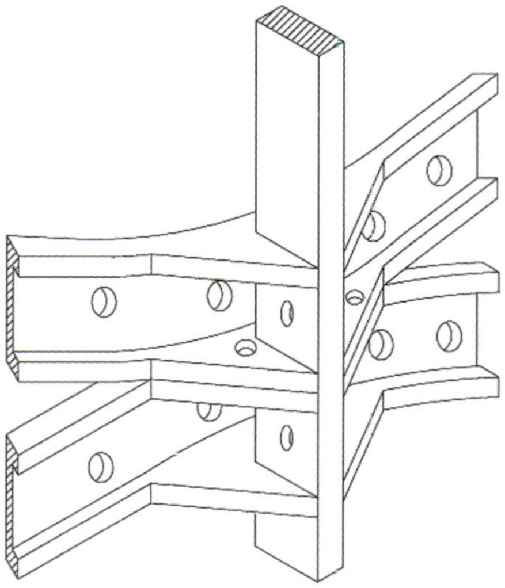

基本铸钢节点
Cast Steel Node

Structural Design

In view of Yingbin Tower's construction characteristics, the use of cast steel components can achieve the smooth transition of its waves. The cast steel nodes are prefabricated in the factory, therefore there should be a small amount of field welding. The use of cast steel nodes can meet the requirement to complete the construction in short-term, saving human, material and construction time.

The cast steel components are prefabricated in the factory, with machine methods and hot dip galvanized. The smooth surface of treated cast steel should make the Yingbin Tower more like a work of art.

The component of cast steel nodes are German standard G20Mn5. This project has a total of 624 nodes, each weighing about 0.7 tons.

机械加工
Machine Methods

节点分析与试验

为了验证铸钢节点的受力性能，根据Midas Gen ver.730建筑结构通用有限元软件的分析结果，选择最不利的静力工况，并提取该工况下节点内力。利用大型通用有限元分析软件Ansys14.0对铸钢节点进行静力分析，杆件应力均小于铸钢材料屈服强度（390MPa），满足强度要求。

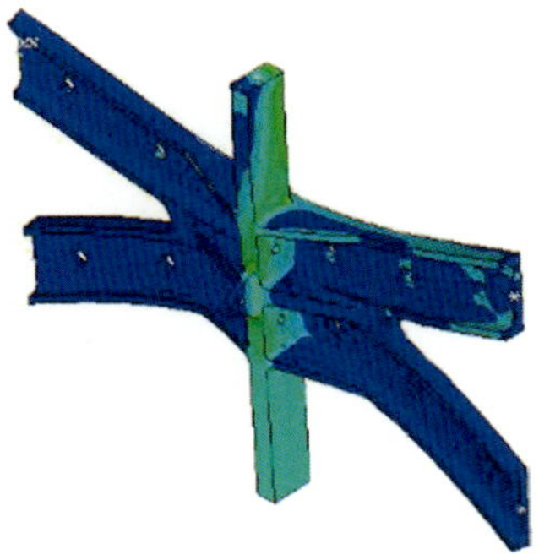

有限元分析结果
Result of Analysis

Node Analysis and Test

In order to understand the mechanical property of cast steel connections, Midas Gen ver.730 finite element software was used to select the most unfavorable static conditions and to pick up the internal force. According to the general finite element analysis software Ansys14.0's static analysis results, bar stress is lower than the cast steel yield strength (390MPa), which meets the strength requirements.

节点试验现场 Scene of testing site

在有限元分析结果的基础上，通过对铸钢节点进行足尺模型试验，明确了铸钢节点的受力性能。试验结果表明，在设计荷载作用下，节点域应力均处于弹性阶段，安全度较高，符合设计与使用要求。

On the basis of the results of the finite element analysis, the full scale model tests of the cast steel connections were conducted. The result clearly shows the mechanical property of cast steel connections, and it demonstrates that under the design load the connection domain stress is within the elastic level. It have high safety factor and is satisfied with the design and use requirements.

建造、施工现场
Construction Site

照明设计 Illumination Design

德国戴水道设计公司 Atelier Dreiseitl GmbH
央美光成（北京）建筑设计有限公司
Lighting City （Beijing）Architectural Design Co., Ltd

摄影：高寒
Photographer: Gao Han

整体照明设计 The Overall Lighting Design

以简洁、清新、明朗设计风格作为文化中心景观照明设计的准则

平时照明在满足功能性照明的前提下，打造多样丰富优美的城市空间和景观环境。

节假日对平时照明进行景观性照明补充，突出温馨、大气、赏心悦目的环境氛围。

重大节日是在节假日基础上的强化，景区呈现出一派欢快、热烈的景象，提升城市的活力与魅力。

In Cultural Central Park, the design styles of concise, fresh and clear have defined as the principles of landscape lighting design.

While on the condition that the daily lighting is supplied functionally, it creates various, plentiful and beautiful urban space and landscape environment.

To build a powerful, pleasant and warm environment on festival and holiday, it needs to increase landscape lighting. And for some major festivals the landscape lighting is enhanced. So that it gives an effect of scenic spot is being happy and cheerful and city is being active and charming.

整个文化中心景观照明不是区域性高亮照明区域，在城市广场中的地位应是使人放松的，安静的，局部点亮的城市绿色漫步空间，亮度不宜过亮，以满足安全功能为主。

Landscape lighting is not the main focus of the whole cultural centre, it should make people feel at ease and quiet in the city square within an urban green walking space. Lighting levels should not be too bright under safety considerations.

步道夜景
Promenade Night View

同时建筑物的照明基调应该依靠投光灯及建筑物的内光外透体现文化中心建筑的高雅品质，利用面性照明体现石材表面的整体统一性，力图做到品质照明，质量照明。

力图使文化中心在今后喧嚣的都市空间中成为市民舒适的休闲场所，同时为今后周边发展高密度城区提供优雅的高品质照明空间。

To emphasize the elegant quality of the cultural center, the buildings are lighted up by tall projection lamps, and the indoor illumination is reflecting the quality of the buildings. The holistic illumination emphasizes the individual components as well as their high quality materials.

The future aim of the cultural center is to become a comfortable recreational area for residents in the noisy urban space. In the meantime, it provides elegant and high-quality lighting space for the surrounding high density urban area in future.

整体照明 Overall Light

中心湖喷泉
Centeral Lake Fountain

北京中科水景科技有限公司
Beijing sino- Academy of Sciences waterscape Technology Co., Ltd.

中心湖犹如文化中心的“心脏”。湖中的喷泉水秀景观与各式建筑相映成趣，飞舞雀跃的喷泉水线随着音乐的跳动，正是文化中心永恒的脉搏，生生不息。

本喷泉工程位于文化中心区域规划的“一湖，一轴，三区”的中心湖内。湖面面积约为9.74公顷。音乐喷泉位于整个规划区域的中心，起到画龙点睛的作用，是文化中心区域的一个景观亮点。中心湖犹如文化中心的“心脏”。

中心湖喷泉总体设计理念为“天圆地方”，中国传统的建筑空间多为“方正”，即用人的视觉、听觉等多方面体验赋予看似无序的自然山水以结构性的秩序，从而使自然空间升华为人文的可把握的自然之精华。方圆相济，为天地，为刚柔，阴阳维让，一合自然，水蕴天成呈无极之象，周而复始，生生不息。

Center lake is like the "heart" of the cultural center. The landscape of center lake and the various buildings exist side by side,offering visitors picturesque views. The streams of fountain dancing to the beat of music, just like the eternal pulse of cultural center which never ends.

"One Lake, One axis, Three District" are planned in the site. This fountain is located in the center of this "One Lake". The lake are is approximately of 9.74 hectares. Musical fountain is located in the center of the entire planning area, as a finishing touch, and a highlights of the cultural center of the region. Center lake is like the "heart" of the cultural center.

The design concept of the central lake fountain comes from the traditional Chinese conception of the universe, "the dome-like heaven embraces the vast earth". Traditional Chinese architectural space is usually upright and foursquare, which uses human sense like vision and hearing to extract structural order from the nature that appears random and orderless, thus distill the essence from natural space. In this conception, square and circle complete each other and form sky and earth. Yin and Yang respect each other and coexist in harmony like nature. The accumulation of water shows a timeless image that everything goes over and over and forms the circle of life.

摄影：刘东
Photographer: Liu Dong

水秀表演 Water Shows

日常表演
Daily Performance

SOMERSET

夜景表演
Night Scene Performance

水型设计

整体音乐喷泉平面，南北宽192米，东西长170米，中心大圆直径80米，小圆直径40米，弧线长156米，与大剧院顶面投影圆重合。平面造型方正，形体大气，造景元素布局方圆结合。本方案以“河海交响曲”为设计主题，展示天津市大气、和谐、现代的靓丽形象。

音乐喷泉水型总体布局为矩阵，喷泉水型的布置形式为点阵、环形、弧线形三种形式，6种喷射装置。层出不穷的造型组合形成错落有致的景观效果。

Water Design

The musical fountain stretches 192meters from north to south and 170 meters from east to west. The great circle in center is 80 meters in diameter, and the small one is 40 meters in diameter. the ellipsis of 156 meters is overlapping with the circle projection of the rooftop of the theater. Based on a rectangular grid pattern, its layout is well combined with the landscape elements. considered as the main theme of the design, the 'Haihe River Symphony' displays the image of generosity, harmoniousness, mordernness of Tianjin city.

The overall layout of musical fountain is based on a matrix pattern. The six kinds of injection equipment should make different forms of appearances, such as lattice, ring, arc-shaped. These organized forms should display the wonderful landscape effect.

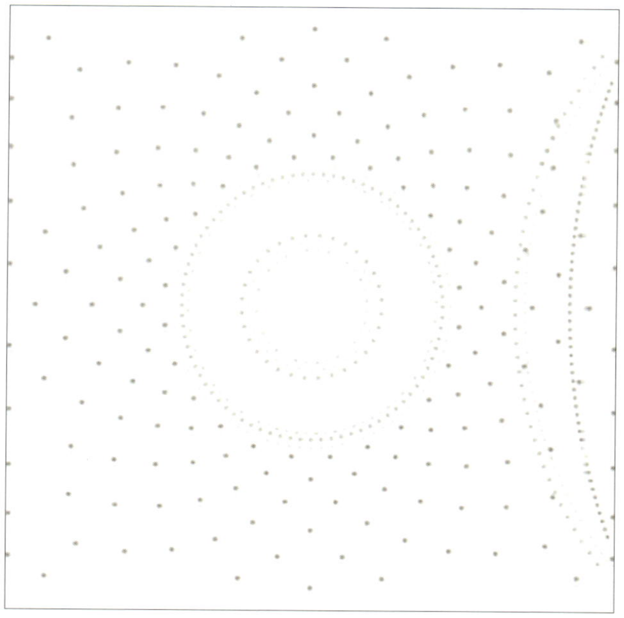

水型设计平面图
Plan for Water Design

水型设计效果图
Rendering for the Water Design

结构设计

本项目中喷泉设备安装采用钢制漂浮升降式平台。喷泉表演时，喷头露出水面，喷泉不表演时，喷泉设备全部沉入水下，中心湖是一个平静的水面，从而保证文化中心整体景观效果。在冬季停运后，将平台降至结冰层厚度以下，保护设备安全越冬。

电控设计

音乐喷泉表演控制软件是音乐喷泉控制系统的核心。本项目软件设计具有手动、程序、实时音控、预编辑音乐控制等多种控制方式，保证长期运行的稳定性。

Structural Design

The project uses steel floating lifting platform for the equipment installation. In show time, only the spray heads go out of the water to give the performance while all the other time the central lake will remain peaceful with the whole device kept underwater. During off work seasons like winter, the platform goes down below the ice layer to ensure the device safety.

Electrical Control Design

The core of the musical fountain control system is the software. It has multiple control types like manual, procedure, real-time sound control, pre-edit music control, etc. The diversity of control methods guarantees the long-run stability of the system.

标识设计 Sign Design

深圳广田装饰集团股份有限公司
Shenzhen Guangtian Decoration Group Co., Ltd.

友谊路
Youyi Road
室内停车场
Indoor Parking
大巴停车场
Bus Parking
公交总站
Bus Terminal
天津自然

城市属于你
绿茵属于你！
City belongs to you,
are you green!

大巴停车
Bus Parking
公交总站
Bus Termina

禁止游泳
No swimming

雕塑创作 Sculpture Creation

作品名称：水中月

Sculpture Name: Moon on the Water

作者：朱尚熹

Sculptor: Zhu Shangxi

摄影：刘东、朱尚熹
Photographer: Liu Dong, Zhu Shangxi

作品名称：母与子

Sculpture Name: Mother and Sor

作者：韩美林

Sculptor: Han Meilin

摄影：刘东
Photographer: Liu Dong

后记

文化是城市的“灵魂”，城市的文化资源、文化氛围和文化发展水平，在很大程度上决定着城市的活力和竞争力。为适应天津经济社会的快速发展，体现城市发展定位，完善城市文化服务功能，提升城市形象，满足人民群众日益增长的文化需要，天津市委、市政府决策规划建设了天津文化中心。从2008年开始，历经三年半的悉心磨砺，“天津文化中心”终于从一系列理念和决策蝉变为一张张规划设计图纸，从精挑细选的一砖一瓦羽化为一栋栋精美的建筑、雅致舒适的环境与大气恢弘的整体城市空间。

天津文化中心以“文化、人本、生态”为主题，在整体规划、理性设计、可持续设计、地域文化特色等方面努力实践，形成以文化为主导的、复合功能的城市核心公共空间与文化场所，集公益文化场馆、城市公园、市民休闲中心、青少年活动场所为一体，成为天津城市高雅艺术的展示中心、文化艺术的普及中心，成为靓丽的城市客厅、和谐的市民乐园，为增强城市文化软实力、丰富精神文化生活、提升市民文化素养发挥作用。期间，国内外40余家著名规划设计团队、百余位院士大师及其他专家学者积极参与，殚精竭虑、出谋划策，十余家施工建设单位日夜奋战、精益求精，各级管理人员尽心尽力、鞠躬尽瘁，市民与社会各界高度关注、全力支持，点滴心血汇成了千余个奋力拼搏的日夜、千余个慎思比选的设计方案和千余个紧张高效的协调统筹会议，而今天这一切都演变成为天津文化中心使用者脸上的盈盈笑意。

将天津文化中心规划设计的过程与成果全面编纂为上下两卷，目的在于忠实记录规划与设计的理念、构思与经验，探讨城市文化空间与建筑设计的内涵和发展趋势，以供在后续文化项目建设实践之中进行分析与总结。天津文化中心投入使用短短数月，已有百余万人次光临这一城市会客厅，来体验感受它的细节、功能与整体气氛。时光前行，快速发展的天津将吸引更多海内外各界人士，钟情于这片神奇的热土，关注天津、共谋发展。天津文化中心也将成为完善城市功能、提升城市面貌、弘扬城市文化、促进城市发展的强有力的城市“心脏”。

2012年10月

Postscript

Culture is the soul of a city. The vitality and competitiveness of a city depend on its cultural resource, cultural atmosphere and the level of its cultural development to a great extent. In order to adapt to the rapid economic and social development of Tianjin, to reflect the development orientation of the city, to improve its cultural service function, to promote its city image, and to meet the growing cultural demands of its citizen, Tianjin Municipal Party Committee and Tianjin Municipal Government decided to carry out the planning and construction of Tianjin Cultural Center. After three and half years of dedicated works from 2008, Tianjin Cultural Center finally becomes pieces of planning and design drawings from a series of ideas and decisions, and becomes fine buildings with elegant and comfortable environment and a whole magnificent urban space from carefully selected a single brick and tile.

Tianjin Cultural Centre has been built upon the spirits of "culture, people, and ecology". Great efforts have been made on the practice of overall planning, rational design, sustainable design, regional cultural features and so on, in order to form a city core and cultural place led by culture and with compound functions. Putting public cultural venues, a city park, a civic recreation center, a youth activity centre together as a whole, Tianjin Cultural Centre has become an elegant art exhibition center, a popular center of culture and art, a beautiful city living room, and a harmonious civic paradise, enhancing the cultural soft power of the city, enriching its spiritual and cultural life, and raising the cultural accomplishment of its citizens. Meanwhile, more than 40 well-known design teams from the country and abroad and more than 100 experts and scholars were involved in this project, racking their brains and giving their advices. And more than 10 construction units worked hard days and nights, pursuing the excellence; managers at all levels made dedicated efforts and tried their very best; citizens and the whole society attached great attention and gave full supports. Drops of efforts made up more than 1000 days and nights of hard work, more than 1000 optional design schemes, and more than 1000 coordination meetings, which all finally become the smiles of the users of the Tianjin Cultural Center today.

The achievement of the planning and design of Tianjin Cultural Center has been completely compiled into two volumes, which record the ideas, concepts, and experiences of its planning and design, the purpose is faithfully recording the process of its rational evolution, optimization and finalization, and which probe the connotation and the developing trend of urban cultural space and architectural design, in order to make thorough analysis and summary for the further practice of cultural construction projects. Since the Tianjin Cultural Center was put into use, in just a few months, over millions of people have visited this city living room, experiencing its details, functions and entire atmosphere. As time going, the rapid development of Tianjin will be more widely to attract more people from all walks of life at home and abroad to come to Tianjin for common developments. We look forwards that Tianjin Cultural Center could become a strong heart of the city to improve its functions, to enhance its image, to promote its culture and to stimulate its development.

Sep 2012

英文翻译：窦强
English Interpret：Dou Qiang